The Logician and the Engineer

0101100101100101100010

Logic: another thing that penguins aren't very good at.

The Logician and the Engineer

0101100101100101100010

How George Boole and Claude Shannon Created the Information Age

PAUL J. NAHIN

PRINCETON UNIVERSITY PRESS
PRINCETON AND OXFORD

Published by Princeton University Press, 41 William Street, Princeton, New Jersey 08540

In the United Kingdom: Princeton University Press, 6 Oxford Street, Woodstock, Oxfordshire OX20 1TR

press.princeton.edu

Third printing, and first paperback printing, 2017

Paperback ISBN 978-0-691-17600-0

The Library of Congress has cataloged the cloth edition of this book as follows:

Nahin, Paul J.

The logician and the engineer : how George Boole and Claude Shannon created the information age / Paul J. Nahin. pages cm Includes bibliographical references and index.

ISBN 978-0-691-15100-7

1. Boole, George, 1815–1864. 2. Shannon, Claude Elwood, 1916–2001. 3. Logicians—Great Britain—Biography. 4. Electrical engineers—United States—Biography. 5. Computer logic. I. Title.

QA76.9.L63N34 2012

510.92'2-dc23

[B]

2011053127

British Library Cataloging-in-Publication Data is available

This book has been composed in ITC New Baskerville

Printed on acid-free paper. ∞

Typeset by S R Nova Pvt Ltd, Bangalore, India

Printed in the United States of America

3 5 7 9 10 8 6 4

For Patricia Ann

who fifty years ago changed the world of a geeky denizen of

the monastic Caltech graduate dorms

Contents

Author's Note to the Paperback Edition

010110010110010110010

A HARDCOVER BOOK GOING TO a corrected paperback edition is always a wonderful opportunity for the author. That's because—at last—the moment has arrived when all the sleep-depriving screw-ups can be banished. Every book ever written–including any novel by Charles Dickens, Stephen King, and Tom Clancy–has a typo or two lurking *somewhere*. Books are created by humans, and humans are fallible. Numerous readers have taught me that lesson, many times, during the more than thirty-year period in which I have been writing books. Still, to be honest, I would not be surprised if at least one more typo has managed to escape detection.

Well, enough of the defeatist self-tortures! Let me simply tell you that all the annoying typos of which I am aware have been corrected in the following pages. There was also one somewhat more significant mistake that was simply too big to quietly correct in the text. At the bottom of page 54 is a directive referring the reader to this very note. Where that directive is, there was, in the hardcover edition, an incorrect equation. In the box below is what I should have written, right after the last line in the text reading "by inspection,"

$$TC + CJ = C(T + J) = 1$$

and so $C = 1$ and $T + J = 1$. The $T + J = 1$ result says that either $T = 1$ and $J = 0$, or that $T = 0$ and $J = 1$. The formal possibility of $T = J = 1$ is excluded because we just showed that $C = 1$, and we can have only one more variable equal to 1. To determine which of T and J is the remaining variable equal to 1 (that is, is the solution $C = T = 1$, or is it $C = J = 1$?), we try both possibilities. First, set

$C = T = 1$ and set all other variables equal to 0. Then, the third factor of (4.5.6), $(T + C)$, has both statements true (1), which is forbidden by the problem statement. So, let's try $C = J = 1$ and set all other variables equal to 0. Then (4.5.6) becomes

$$(1 + 0)(0 + 0)(0 + 1)(0 + 1)(0 + 1) = 0$$

and so we see that every factor but one on the left-hand-side has one true statement (1) and one false statement (0), and the remaining factor (the second one) has two false statements. This satisfies the problem statement.

As your eyes exit this box please continue reading at the top of page 55.

I gratefully thank Princeton University Press for the opportunity to make these corrections.

Paul J. Nahin
Exeter, New Hampshire
July, 2016

Preface

010110010110010110010

THIS BOOK IS ABOUT an amazing intellectual "collaboration" between two men who never met. The Englishman George Boole lived his entire life within the nineteenth century, while the American Claude Shannon was born in the twentieth and died at the beginning of the twenty-first. Boole, of course, never knew Shannon, but he was one of Shannon's heroes. It is because of Shannon that Boole is rightfully famous today, but it is because of Boole that Shannon first gained the attention of the scientific community.

What makes the cross-time relationship of these two remarkable men particularly interesting is that Boole was a pure mathematician, a man who lived in the rarefied, abstract world of the academic, while Shannon was primarily a practical, "get your hands dirty" electrical engineer. Despite this extreme difference in their worldviews, it is simply impossible to think of one of these men without thinking of the other. So many of the well-known scientific theories of our day are attached—rightfully or not—to a single name (it is *Einstein's* theory of special and general relativity, it is *Newton's* theory of gravity, it is *Maxwell's* theory of electrodynamics, it is *Darwin's* theory of evolution, it is *Schrödinger's* theory of quantum wave mechanics, it is *Heisenberg's* theory of quantum matrix mechanics, and so on), but when one hears of *Boolean* algebra one immediately thinks also of *Shannon's* switching theory. And vice versa. The two names are intimately entangled.

Later in his life Shannon's name did become uniquely attached to the new science of information theory, but even then you'll see as you read this book how the mathematics of information theory—probability theory—was a deep, parallel interest of Boole's as well.

What Boole and Shannon created, together, even though separated by nearly a century, was without exaggeration nothing less than the fundamental foundation for our modern world of computers and information technology. Bill Gates, the late Steve Jobs, and other present-day *business* geniuses are the people most commonly thought of when the world of computer science is discussed in the popular press, but knowledgeable students of history know who were the real technical minds behind it all—Boole and Shannon (and Shannon's friend, the English genius Alan Turing, who appears in the following pages, too).

Read this book and you'll understand why.

The Logician and the Engineer

1

What You Need To Know to Read This Book

0 1 0 1 1 0 0 1 0 1 1 0 0 1 0 1 1 0 0 1 0

If a little knowledge is dangerous, where is the man who has so much as to be out of danger?
—Thomas Huxley (1877)

Claude Shannon's very technical understanding of information . . . is boring—it's dry.
—James Gleick, in a 2011 interview about his book *The Information*, expressing a view common among those who think glib cocktail conversation is equivalent to analytic reasoning and true understanding

TO READ THIS BOOK you don't have to be an electronics genius, a computer geek, or a quantum mechanics whiz. But that doesn't mean I'm assuming you are a high school dropout, either. I will, in fact, be assuming some knowledge of mathematics and electrical physics and an appreciation for the value of analytical reasoning—but no more than a technically minded college-prep high school junior or senior would have. In particular, the math level is that of algebra including knowing how matrices multiply. The electrical background is simple: knowing (1) that electricity comes in two polarities (positive and negative) and that electrical charges of like polarity repel and of opposite polarity attract; and (2) understanding Ohm's law for resistors (that the voltage drop across a resistor in *volts* is the current through the resistor in *amperes* times the resistance in *ohms*) and the circuit laws of Kirchhoff (that the sum of the voltage drops around any closed loop is zero, which is an expression of the conservation of energy; the sum of all the currents into any node is zero, which is an expression

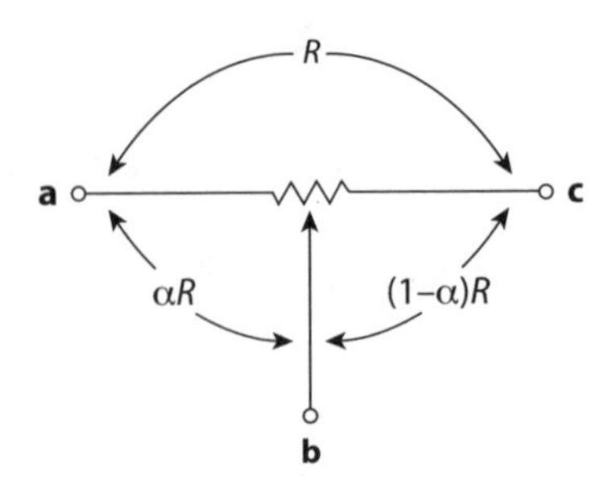

Figure 1.1. A potentiometer is a three-terminal circuit element.

of the conservation of electric charge). *No knowledge of electronics is required.*

That's it.

That's pretty brief, too, and so here is an example of the level of math/physics I am going to assume on your part. As you read through what follows, ask yourself if, at each step, you understand the reasoning. If you can say "yes" all the way through, then you can be sure there will be nothing in this book beyond your grasp. My example is from a minor classic in electrical circuit theory, a problem studied in a 1956 paper coauthored by the mathematical electrical engineer Claude Shannon. (I specifically mention Shannon here because—besides being mentioned by Gleick—he is a central character in this book.) That paper opens with the following amusing words

> As part of a computer, a rheostat having a resistance that was a concave upward function of the shaft angle was needed. After many attempts to approximate it with networks of linearly wound potentiometers and fixed resistors, it became apparent that either it was impossible or that we were singularly inept network designers. Rather than accept the latter alternative, we have proved the [impossibility of such a network].[1]

Now, first of all, a couple of explanations.

A *potentiometer* (or *rheostat*, a rather old-fashioned word) is simply a *variable resistor*, with the variation produced by rotating an attached shaft; a *linear* potentiometer is one whose variable resistance is a linear function of the shaft rotation angle. Figure 1.1 shows a potentiometer

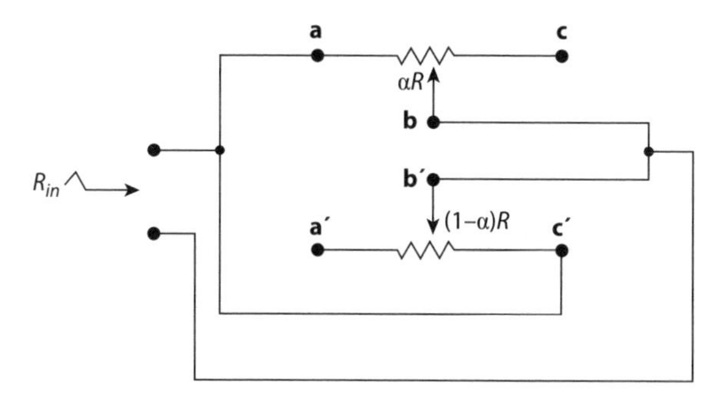

Figure 1.2. Two potentiometers in parallel.

as a three-terminal element, with the fixed resistance R available between the terminals marked **a** and **c**, the variable resistance αR available between the terminals **a** and **b** (**b** is the output terminal on a sliding contact—called the *wiper arm*—connected to a rotatable shaft which is, in turn, connected to a control panel knob), and the variable resistance $(1-\alpha)\ R$ available between terminals **b** and **c**. We imagine that the non-negative parameter α is directly proportional to the shaft rotation angle, with $\alpha = 0$ representing the shaft turned all the way counterclockwise (**b** to the extreme left) and $\alpha = 1$ representing the shaft turned all the way clockwise (**b** to the extreme right).

A *concave upward* function is one that, to invoke some easy imagery, is a curve that can "hold water." (Take an advance look at Figure 1.3 for an example of the opposite case, that is, a curve that can *not* "hold water.") What Shannon and his coauthor proved, then, is that such a "water-holding" function can *not* be realized by *any* combination of linear potentiometers and fixed resistors. Their general proof is rather subtle, and is far more sophisticated than anything we'll do in this book. But for any given, specific circuit, we can confirm their result by direct calculation.

So, consider Figure 1.2, which shows two potentiometers wired in parallel. The two potentiometers are mechanically "ganged," which means that the same shaft simultaneously varies the two wiper terminals (**b** and **b**′). The two potentiometers have the same fixed resistance R (between **a** and **c**, and between **a**′ and **c**′), but are electrically connected so that the parallel resistances are αR and $(1-\alpha)R$. Since

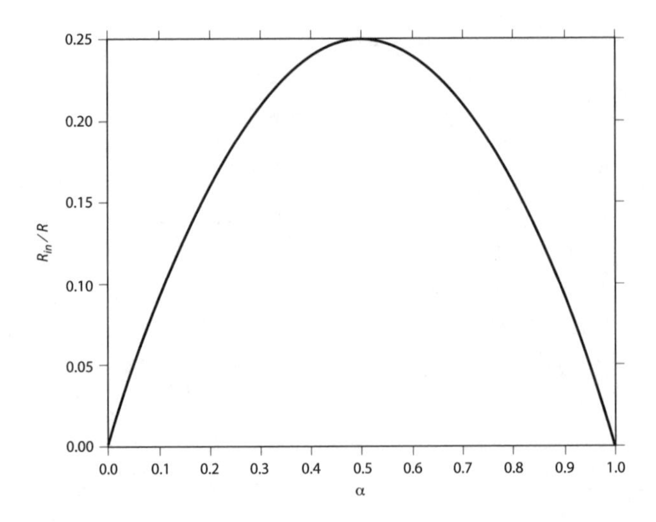

Figure 1.3. A resistance function that cannot 'hold water'.

elementary circuit theory (stuff I'm expecting you to know at the outset) tells us that the equivalent resistance of two parallel resistances is their product divided by their sum, then we immediately have the resistance of the circuit in Figure 1.2 as

$$R_{in} = \frac{\alpha R(1-\alpha)R}{\alpha R + (1-\alpha)R} = \frac{\alpha(1-\alpha)R^2}{R}$$

or,

$$\frac{R_{in}}{R} = \alpha(1-\alpha).$$

Figure 1.3 shows a plot of $\frac{R_{in}}{R}$ for $0 \le \alpha \le 1$, and it is indeed a concave *downward* function (a parabola, in fact), in agreement with the general result of Shannon and Hagelbarger.

Okay, did all that make sense to you? It won't to everyone. Shannon was a jazz buff and, in a 1952 talk on analytic creativity he gave at Bell Labs, he repeated jazz legend Fats Waller's famous remark about who could play swing music: "Either you got it or you ain't." If all the above *did* make sense to you, then "you got it" for what it takes to read this book.

NOTES AND REFERENCES

1. C. E. Shannon and D. W. Hagelbarger, "Concavity of Resistance Functions," *Journal of Applied Physics*, January 1956, pp. 42–43. When the authors write of their problem being inspired by the design of "part of a computer," I'm assuming that means they were thinking of an *analog* computer and not a digital machine. Shannon had, in fact, worked for a time (1936–38) on the famous "differential analyzer" at MIT (where he received, at the same commencement in the spring of 1940, a master's in electrical engineering and a doctorate in mathematics). The analyzer was the most advanced electro-mechanical analog computer in the world. Analog computers are generally not thought to be as "sexy" as digital computers, but I think that is quite wrong. To see what an analog computer can do with a pretty complicated math problem, see my book *Number-Crunching*, Princeton University Press 2011, pp. 253–259. An alternative proof of the Shannon-Hagelbarger theorem is given in the *Journal of Applied Physics* by H. M. Melvin, June 1956, pp. 658–659.

2

Introduction

0101100101100101100100

"Contrariwise," continued Tweedledee, "if it was so, it might be; and if it were so, it would be; but as it isn't, it ain't. That's logic."

—Lewis Carroll, *Through the Looking-Glass* (1872)

IN 1859 THE ENGLISH NATURALIST Charles Darwin (1809–1882) published his *On the Origin of Species*, a book that revolutionized how humans view their place in the world. Just five years earlier a fellow countryman, the mathematician George Boole (1815–1864), had published his *An Investigation of the Laws of Thought*, a book that would have an equally huge impact on humanity. Even earlier, in fact, Boole had published his *Mathematical Analysis of Logic* (1847), which was, in essence, a first draft of *Laws of Thought*. The importance of Boole's work was not as much appreciated at the time as was Darwin's, however, because it wasn't enough by itself to have immediate influence. It required one additional contribution, one that didn't come about until decades later, in 1938, with the work of the American electrical engineer and mathematician Claude Shannon (1916–2001). That was the year Shannon published a famous paper (based on his MIT master's thesis) on how to implement Boole's mathematics in the form of electrical relay switching circuits. Together, Boole's and Shannon's ideas ushered in the digital age.

Boole's mathematics, the basis for what is now called *Boolean algebra* (although it is different in some significant details from what Boole actually wrote), is the subject of this book. It is also called *mathematical logic*, and today, because of Shannon, it is a routine analytical tool of the logic-design engineers who create the electronic circuitry that each of us now can't live without, from our computers to our automobiles to our home appliances. Boolean algebra is *not* traditional or classical Aristotelian logic, a subject generally taught in college by the philosophy department. Boolean algebra, by contrast, is generally in the hands of electrical engineering professors and/or the mathematics faculty (although, of course, philosophers are familiar with it, too)[1]

Aristotelian logic was created to develop correct thinking in those who would serve in the legal profession, whose work often involves the logical deduction of a conclusion (for example, *guilty* or *innocent*) from a set of given conditions (that is, the *evidence*). When one starts from given premises and then arrives at a conclusion logically implied by those premises, Aristotelian logicians say we have constructed a *syllogism*. A syllogism has the general form of a *major* premise and a *minor* premise, linked together, from which inexorably follows a *conclusion*. An example is

Major premise: "All physicists have studied mathematics."
Minor premise: "You are a physicist."
Conclusion: "You have studied mathematics."

A false syllogism is the slightly modified

Major premise: "All physicists have studied mathematics."
Minor premise: "You have studied mathematics."
Conclusion: "You are a physicist."

Whole books, indeed, *libraries* of books, have been written over the centuries on this sort of logical reasoning. This book is not one of them. Boolean algebra is an entirely different form of logical reasoning, as you'll soon see. Indeed, it was eventually realized that the classical syllogism could not be the final step in logical reasoning. Consider, for example, the two premises

"Most people have studied physics."
"Most people have studied mathematics."

Here "most" means "more than half." Classic syllogistic logic doesn't say how to draw a definitive, inexorable conclusion from these two premises. But, in fact, a perfectly valid conclusion *is* possible:

"Some people have studied both mathematics and physics," where "some" means "one or more."

Here's another example of how classical logic can stumble. One of the underlying principles of that subject is that of the *excluded middle*, which postulates that every statement is either true or false. There is no "middle ground." So, consider the statement "I am lying." The principle of the excluded middle says this statement is either true or false. So, suppose it's true. That is, "I am lying" is true, and so I *am* lying. But, since the statement is true, I'm *not* lying. Contradiction! On the other hand, suppose "I am lying" is false. That is, the statement "I am lying" is a lie. But that means the statement is true. Contradiction! True *or* false, we fall into contradiction, and so the principle of the excluded middle seems to have failed us here.

The point of these two examples is simply that classical logic isn't the absolute, final word in logic.

Now, before I dive into Boolean algebra I should be up-front with you and admit that, while Boolean algebra is of immense importance—otherwise, why this book!?—it has its limitations, too. To show you the latter, consider the following two problems. The first is a curious puzzle from *classical* logic for which Boolean algebra can offer us no help. Here is how Lewis Carroll stated what is known today as the "Paradox of the Court," in Part 2 of his brilliantly eccentric book *Symbolic Logic*:

> Protagoras had agreed to train Euathius for the profession of a barrister, on the condition that half his fee should be paid at once, and that the second half should be paid, or not paid, according as Euathius should win, or lose, his first case in Court. After a time, Protagoras, becoming impatient [at Euathius's delay at beginning his career, presumably to avoid paying the second half], brought an action against his pupil, to recover the second half of his fee. It seems that Euathius decided to plead his own case. "Now, if I *win* this action," said Protagoras, "you will have to pay the the money by the decision of the Court: if I *lose* it, you will have to pay by our agreement. Therefore, in any case you must pay it." "On the contrary", retorted Eauthius, if you *win* this action, I shall be released from payment by our agreement: if you *lose* it, I shall be

> released by the decision of the Court. Therefore, in any case, I need not pay the money.[2]

Which man is right?

The appeal of this puzzle for the agile intellect of Lewis Carroll (the pen name for the University of Oxford mathematics teacher Charles Lutwidge Dodgson (1832–1898)) is clear. Dodgson loved the play of words, and of double meanings, a claim made obvious by reading his classics *Alice's Adventures in Wonderland* and *Through the Looking-Glass*. And that is just what the "paradox" of the Court is, a play on words. It is only slightly more subtle than the old schoolboy question of "What happens when an irresistible force meets an unmovable object?," a question reduced to nonsense once it is realized that it is self-contradictory. If there is an irresistible force, then by definition there can be no unmovable object, and vice versa. To claim both, simultaneously, is similar to saying something (for example) is simultaneously dead and alive (or black and white, or, in general, simultaneously possessing any two mutually exclusive properties) and then to be puzzled at a resulting "paradox."

The answer to the Court paradox is that either man may prevail, depending on the declaration by the Court, and the resulting "conundrum" occurs only after that declaration (at which point the puzzle is, in fact, no longer a puzzle). Here's how Lewis Carroll answered the question (you'll notice that Boolean algebra—which was fully available when he wrote—makes no appearance), as well as explaining how timing is crucial.

> The best way out of this Paradox must seem to be to demand an answer to the question "*Which* of the two things, the agreement and the decision of the Court, is to over-ride the other, in case they should come into collision?"
>
> (1) Let us suppose that the *agreement* is to be supreme. In this case, if Protagoras *wins* his action, he *loses* the money; and if he *loses* his action, he *wins* the money.
>
> (2) Let us suppose the *decision of the Court* to be supreme. In this case, if Protagoras *wins* his action, he *wins* the money; and if he *loses* his action, he *loses* the money.

> The *Data* do not enable us to answer this question [of which of the two things is to be supreme]. Protagoras naturally makes one, or the other, supreme, as best suits his purpose and his docile pupil follows his example [but, of course, with the opposite choice].

After writing the above Lewis Carroll goes on to offer additional commentary that nicely explains how the timing of the Court's decision comes into play:

> The right decision of the Court would obviously be *against* Protagoras, seeing that the terms of the agreement were still unfulfilled [because until the Court renders its decision Euathius *has not yet won his first case*!]. And, when that decision has been pronounced, the practical result would be that, if the *agreement* was to be supreme, Euathius would have to pay the money: if the *decision of the Court* was to be supreme, he would be released from payment.

None of the modern presentations of this puzzle that I've seen on the Internet are as penetrating as is this century-old explanation.

For my second illustration of a logic problem for which Boolean algebra is of no use, let me show you an example of a logic problem which—as far as I know—can't be solved by *any* formal system of logic, Aristotelian, Boolean, whatever. This is a logic problem that is, indeed, "solvable," but only by making an argument that at one point goes beyond formal mathematics. That is, some additional human insight is required. My example below is well known among mathematicians, but seems not to be so well known to wider audiences. I'll let you ponder it for a while and, if you get stuck, the answer is in the last note of this chapter. But don't look until you've thought about it for at least a while.

> Two mathematicians (I'll call them **A** and **B**), friends in college who have not seen each other for forty years, meet at a conference. As they have lunch together in a local restaurant, **A** tells **B** that he has three daughters and that the product of their ages (all integers) is 36. **A** then challenges **B** to determine the individual ages. **B** of course replies that the ages are not uniquely determined. So, **A** gives **B** a second clue: the sum of the ages is

> equal to the number of people in the restaurant. **B** looks around and then again replies that the ages are still not determined. So, **A** provides a third hint: his oldest daughter loves to eat bananas. "Ah, ah," shouts **B**, immediately "*now* I have it" and he then promptly and correctly gives the ages.

Okay, now *you* solve this problem in logical reasoning.

So, after presenting the last two examples the obvious question now is: What sort of logic problem *would* suggest the application of Boolean algebra? Consider, then, the following problem that I've called Puzzle 1.

PUZZLE 1

On the table before you are three small boxes, labeled **A**, **B** , and **C**. Inside each box is a colored plastic chip. One chip is red, one is white, and one is blue. You do not know which chip is in which box. Then, you are told that of the next three statements, *exactly one* is true:

(a) box **A** contains the red chip;

(b) box **B** does not contain the red chip;

(c) box **C** does not contain the blue chip.

You do not know which of the three statements is the true one. From all this, determine the color of the chip in each box. Puzzle 1 can be solved by some very careful reasoning that I'll show you at the end of this chapter (but don't peek until you've given it a good try yourself); it can also be solved through the routine application of Boolean algebra.

Now, just to really convince you that Boolean algebra is a most powerful tool, let me ask you to consider the next three puzzles, ones that I feel confident you will *not* be able to solve with simply "some very careful reasoning" or, at least, not until you've expended considerable mental effort. And yet, as we proceed through the book, I'll show you how they too will easily yield to routine Boolean algebraic analysis.[3]

PUZZLE 2

The local truant officer has six boys under suspicion for stealing apples. He knows that only two are actually guilty (but not which two), and so he questions each boy individually.

(a) Harry said, “Charlie and George did it.”

(b) James said, “Donald and Tom did it.”

(c) Donald said, “Tom and Charlie did it.”

(d) George said, “Harry and Charlie did it.”

(e) Charlie said, “Donald and James did it.”

(f) Tom couldn’t be found and didn’t say anything.

(g) Of the five boys interrogated, four of them each correctly named *one* of the guilty.

(h) The remaining boy lied about both of the names he gave.

Who stole the apples?

PUZZLE 3

Alice, Brenda, Cissie, and Doreen competed for a scholarship. “What luck have you had?” someone asked them.

(a) Said Alice: “Cissie was top. Brenda was second.”

(b) Said Brenda: “No, Cissie was second, and Doreen was third.”

(c) Said Cissie: “Doreen was bottom. Alice was second.”

(d) Doreen said nothing.

Each of the three girls who replied made two assertions, of which only one was true. Who won the scholarship? More generally, in what position did each of the four girls finish?

PUZZLE 4

Four hunters, A, B, C, and D, occupied a camp for seven days.

(a) On days when A hunted, B did not.

(b) On days when B hunted, D also hunted, but C did not.

(c) On days when D hunted, A or B hunted.

(d) No two days were identical in who hunted and who didn't.

On how many days did D hunt, and with whom?

Okay, have you solved Puzzle 1? If not, here's how to do it "with reasoning." (By the end of Chapter 4 we'll have solved all four puzzles with the techniques of Boolean algebra.) Since we are told only one of the three statements is true, then we can attack the problem as follows: Take each one of the statements, in turn, as the true one, and *reverse* the other two. If we have selected the correct true statement, then we'll have three true statements. Since there are only three statements in all, we only have to do this three times. For each group of "corrected" three statements we can then see if what they say, collectively, makes sense. So,

Case 1:

Take (a) as true, and (b) and (c) as false. Then, with reversals, we have

(a1) box **A** contains the red chip;

(b1) box **B** contains the red chip;

(c1) box **C** contains the blue chip.

This is, of course, obvious nonsense as (a1) and (b1) cannot both be true.

Case 2:

Take (b) as true, and (a) and (c) as false. Then, with reversals, we have

(a2) box **A** does not contain the red chip;

(b2) box **B** does not contain the red chip;

(c2) box **C** contains the blue chip.

Since box **C** has the blue chip, then the red and white chips are in boxes **A** and **B**. In particular, one of those two boxes must have the red chip, but (a2) and (b2) deny that. Thus, Case 2 is also nonsense.

Case 3:

Take (c) as true, and (a) and (b) as false. Then, with reversals, we have

(a3) box **A** does not contain the red chip;

(b3) box **B** contains the red chip;

(c3) box **C** does not contain the blue chip.

This works. (b3) says **B** has the red chip. That leaves the blue and white chips for **A** and **C**. (c3) says **C** does not have the blue chip, so **C** must have the white chip. Thus, **A** must have the blue chip, which is consistent with (a3).

The author of a well-known science fiction story, in which the narrator is a college math major, opens his tale with the student complaining about his courses.[4] In particular, his class in logic generates the lament, "If it seems to make sense it isn't mathematical logic!" By the time you finish this book I hope you'll reject that sentiment and, instead, agree with me that if mathematical logic is about anything, it *is* about making sense.

Okay, have you solved the "two mathematicians" puzzle? If not, take a look at the final note.[5]

NOTES AND REFERENCES

1. When I was an undergraduate electrical engineering major at Stanford (1958-62), I took the first-year graduate course EE 266 in my senior year. That course ("Digital Computers") was my one and only class in digital combinatorial and sequential logic, and its core mathematics was Boolean algebra. (The previous year I had taken Industrial Engineering 161 which had a programming project on an IBM 650 computer with its famous rotating magnetic drum memory, but that was really an engineering *economics* course, not a logic design course.) When I taught the same material in EE 266 at the University of New Hampshire some years later, at least half of every class

were sophomores—an example of how concepts once thought difficult are now thought to be not so difficult. At the time, I took EE 266 simply because it was fun, with no idea that less than two years later I would be earning my living as a digital logic design engineer. I had that job from mid-1963 to the end of 1965 (see my book *Number-Crunching*, Princeton University Press, 2011 for details) and must have written, or so it then seemed, fifty thousand Boolean equations—much to the occasional irritation, I might add, of my wife, who wondered why I was always scribbling long, alphabetic strings of symbols on paper sheets strewn all over her otherwise immaculate home.

2. Part 1 of *Symbolic Logic* was published in 1896 but Part 2, in which the "Paradox of the Court" appears, remained virtually unknown until 1977. I have taken my quotation of Lewis Carroll's explanation of the paradox from William Warren Bartley III, *Lewis Carroll's Symbolic Logic*, Clarkson N. Potter, 1977, pp. 426, 438.

3. Puzzle 1 is from T. J. Fletcher, "The Solution of Inferential Problems by Boolean Algebra," *Mathematical Gazette*, September 1952, pp. 183–188. Puzzle 2 is from Hubert Phillips, *Heptameron*, Eye & Spottiswoode, 1945. Puzzle 3 is from Hubert Phillips, *Something to Think About*, Max Parrish, 1958 (first published by Penguin in 1945). Phillips provides non-Boolean algebraic solutions in his books. Puzzle 4 is taken from Martin Gardner, *Logic Machines and Diagrams*, University of Chicago Press, 1982 (no solution was given by Gardner, however).

4. Norman Kagan, "Four Brands of Impossible," *The Magazine of Fantasy & Science Fiction*, September 1964.

5. Writing down all the possible ways to form 36 as the product of three integers, we have

Ages			Sum of Ages
2	6	3	11
4	3	3	10
2	2	9	13
4	1	9	14
2	1	18	21
1	6	6	13
3	1	12	16
1	1	36	38

The second clue, about the sum of the ages, would have been sufficient to uniquely determine the ages if the number of people in the restaurant had been 10, 11, 14, 16, 21, or 38, since each of those sums occurs just once. Since that clue was not sufficient, however, then the sum must have been 13, which is the one sum that occurs more than once. That is, the ages are either 2, 2, and 9, or 1, 6, and 6. The last clue, that the *oldest* girl (notice the singular) loves bananas, says there is a *single largest* age, and so the ages must be 2, 2, and 9 (for the sequence of 1, 6, and 6, the third clue would be grammatically incorrect). And so we see that the appearance of fruit in the third clue is nothing more than a colorful fish (that is, a red herring). The crucial word is *oldest*, not *bananas*. If, however, **A** had given as his third clue that the *older* girls (notice the plural) love bananas, then the correct answer would (again, by grammar, not math) be 1, 6, and 6.

3

George Boole and Claude Shannon

Two Mini-Biographies

3.1 THE MATHEMATICIAN[1]

"Oh please, we are playing at lions and we want a good lion who can roar well. Do come and help!"

—The shouts of the adult George Boole's neighborhood youngsters, pleading with the friendly local man who they almost surely didn't realize was a brilliant mathematician in addition to being a "capital lion."

George Boole was born in Lincoln, a town in the north of England, on November 2, 1815. The first of four children born to John (1777–1848) and Mary Ann Boole (1780–1854)—his siblings, a sister and two brothers, all outlived him by decades, with his youngest brother surviving until 1902—he was particularly lucky with his father. While a simple tradesman (a cobbler), he was also a kind, generous, religious man who had a strong interest in both mathematics and the construction of optical instruments. He provided emotional stability and intellectual stimulation, if not wealth, to his family, and George was a devoted son. Even at a young age he enjoyed working alongside his father at optical crafting. One of John's projects was the construction of a telescope and, when it was finished, he placed the following invitation in his shop window:

> Anyone who wishes to observe the works of God in a spirit of reverence is invited to come in and look through my telescope.

George must have been impressed with what he saw of the heavens through that instrument, and perhaps that was one of the influences that propelled him toward his career as a profoundly creative mathematician. A *camera obscura* was another of their joint projects.

John may, in fact, have spent too much time on optics and not enough on repairing shoes. In 1956 Sir Geoffrey Taylor (1886–1975), a mathematical physicist of some renown, a Fellow of the Royal Society, and a grandson of George (his mother was the second of Boole's five daughters), wrote the following: "I have inherited from my grandmother [Boole's eventual wife, Mary Everest (1832–1916)], a box made by John Boole to hold a microscope he had made. Inside the lid is pasted a note in her handwriting [declaring], 'He seems to have been able to do anything well except his own business of managing the shop.'"

A disinterested shopkeeper John may have been, but he did the best he could with limited resources for his children. He taught George geometry and trigonometry, subjects John had found of great aid in his optical studies. George was soon recognized as a highly intelligent boy and was admired by his schoolmates early-on. One of them remembered after Boole's death that "he was not of my class, or indeed of any class; for we had no boy in school equal to him, and perhaps the master was not [either], though he professed to teach him. This George Boole was a sort of prodigy among us, and we looked up to him as a star of the first magnitude." Child prodigies all too often fail—sometimes quite spectacularly—to live-up to their early promise, but Boole would be the happy exception. Indeed, he succeeded in his all-too-short life at exceeding even his childhood friends' great expectations.

As you'll learn in the next mini-biography, the other hero in this book, Claude Shannon, had an extensive, first-class formal education, with graduate degrees in both electrical engineering and mathematics. Boole, on the other hand, was essentially self-taught, with a formal education that stopped at what today would be a junior in high school. Eventually he became a master mathematician (who succeeded—where all others had failed—in merging algebra with logic), one held in the highest esteem by talented, highly educated men who had graduated from Cambridge and Oxford. If such a thing happened today, it would be as if a precocious but impoverished student dropped out of high school in his junior year and then, through years of intense self-study, discovered a proof of the infinity of the twin primes (believed since Euclid's day, but still unproved to this day).

His formal education began at age seven in a primary school, and then briefly at a commercial school concerned with business topics, and he quickly absorbed all that they had to offer. He learned Latin from a local bookseller, and taught himself Greek, French, and German from borrowed books. All of this linguistic study was in preparation for what Boole then thought would be his life's work in the Church of England as a clergyman.

In 1831, at age sixteen, Boole left school to become an assistant teacher, of Latin and mathematics at a small Wesleyan boarding school in Doncaster, forty miles from Lincoln. One of the boarders remembered Boole as follows (in an 1884 letter to Boole's widow):

> During his residence in the school Mr. Boole was much respected for his attainments and for the conscientious discharge of his duty. But when the fact became known that he was a Unitarian [he had grown unhappy with conventional Christian doctrine through his extensive reading of other cultures, and there would be no Holy Orders for Boole!], read mathematics on Sunday and even did problems in Chapel, it marred both his happiness and his usefulness. Complaints were made against him by [families of the boys who reported Boole's "sins"]. The boys prayed for his conversion in their prayer meetings, and this was one reason why [the Headmaster] told him it was desirable his chief assistant should be a Wesleyan.

Boole lost his job at the school in 1833, but perhaps not just for doing math in Chapel. The same boarder also remembered that "with the vast majority of boys, who have no application and require drilling again and again in the same subject, he was the worst teacher I ever met with. Instead of explaining he lost his temper. . . . This was the second cause of his leaving."

His two years in Doncaster were important ones for Boole, as it was there that he switched his intense self-study from languages to mathematics. After a day of being a bad teacher to dull boys, he would use the evening to plow through the pages of what he felt to be an indifferent French book on differential calculus (Lacroix's *Calcul différentiel*). He later proclaimed that effort to have been mostly a waste of time, but it at least got him to the point where he could later read Lagrange's *Calcul des fonctions* and *Mécanique analytique*, Laplace's *Mécanique céleste*, Newton's *Principia*, and Poisson's *Traité de mécanique*. Remember, all this was done by himself, with no fellow classmates to work with in a formal course led by an instructor. As Boole later explained to a friend, he managed it all by sheer force of will, just reading and re-reading, over and over, until he understood.

After leaving Doncaster, Boole took up a similar teaching post at a school in Waddington, a village much closer to Lincoln and his aging parents. The growing dependence of his parents on his support, however, made even the four miles between Lincoln and Waddington burdensome, and Boole eventually found it necessary to move back to Lincoln to start his own day school in 1835. The year before, the Mechanics Institute (offering adult education for the working class) had been founded in Lincoln and the president, a local squire, deposited the publications of the Royal Society in the institute's reading room. Fortuitously, Boole's father was made curator of the institute—a 'family connection' that finally worked for his son!—and George, while still in Waddington, had ready access to the reading room and to all its mathematical treasures.

The arrangement in Lincoln lasted until 1838 when the authorities in Waddington asked George to return, to replace the previous headmaster who had died. He accepted the offer and took his entire family with him in the move, as he was now the sole provider. Apparently it was a good move, as his financial affairs improved dramatically; just two years later, in the summer of 1840, Boole was able to purchase property back in Lincoln to start his own school once more. So once again he moved his entire family, and he remained in Lincoln for the next nine years.

All during these back-and-forth years, Boole's studies of mathematics had continued. In 1838, while still in Waddington, he finally started to put his own original work on paper, writing his first essay, "On Certain Theorems in the Calculus of Variations."[2] He followed it with one bearing the imposing title of "Researches on the Theory of Analytical Transformations, with a Special Application to the Reduction of the

General Equation of the Second Order." So far, however, the only light-of-day those two works had seen was that of the rays making their way through the windows of Boole's study. Who would actually *publish* the writings of an obscure day school proprietor who had never even gone to university, to say nothing of not having graduated with a degree?

Here Boole's fortune received a dramatic, almost heaven-sent blessing. It just so happened that at about the same time Boole was faced with his "where do I send my papers?" question, a new math journal was looking for manuscripts that might not be accepted by the established journals, perhaps (for example) for being too controversial.[3] This was the *Cambridge Mathematical Journal*, which had begun publication in October 1837. From the very start it published papers from such talented people as Augustus De Morgan, Arthur Cayley, James Sylvester, and George Stokes, all mathematicians whose names are well known today. Perhaps, in fact, it was the appearance of this new journal that caused Boole to begin writing.

The young editor of the *Journal*, less than three years older than Boole, was the Scottish mathematician Duncan F. Gregory (1813–1844), who had received a magnificent education that Boole could only have dreamed about. The youngest son of a professor of medicine at King's College in Aberdeen, he had first attended Edinburgh Academy (where James Clerk Maxwell would later study), then was sent off to a private academy in Geneva, then brought back to Edinburgh University and, finally, finished his studies at Newton's school, Trinity College, Cambridge. He graduated from Trinity with a B.A. in 1837 as an impressive fifth wrangler in the famously grueling Mathematical Honors (or Tripos) examination (later, in 1841, he topped it all off with a Cambridge Master's degree).

Despite the disparity in their social backgrounds, there was not the slightest bit of snobbish elitism in Gregory toward Boole. Indeed, without Gregory's almost incredibly generous aid to Boole it is not unreasonable to imagine that Boole's spirit would have been crushed right at the start. Instead, he found a sympathetic reader, one who recognized both the genius and the faults of Boole's initial submission (his second written paper, "Theory of Analytical Transformations") and offered gentle suggestions on how Boole could improve his writing style. Apparently Boole traveled to Trinity College and received those suggestions in person because, in a letter dated November 4, 1839, Gregory in Cambridge wrote to Boole to say, "You spoke *when I saw you here* [my emphasis] of some investigations in the Calculus of Variations, which you were inclined to publish. If you still desire to do so I shall be happy to give them a place in the journal." And so it in fact happened, with Boole's second written paper appearing in print in February 1840, and his first written paper prominently appearing in the journal just three months later (along, as well, with yet another paper, "On the Investigation of Linear Differential Equations with Constant Coefficients"). Those initial papers opened the floodgates, and Boole would publish numerous more papers in Gregory's journal.

Because of his success in achieving publication for his mathematics in a Cambridge-based journal, it isn't surprising that Boole began to think of attending Cambridge University with the goal of earning a degree. He discussed this possibility with Gregory who, while not actually discouraging the then twenty four year-old Boole, was quite candid (in a letter written in March 1840) about the difficulties that such a move would entail. Gregory had some secondhand "experience" at the possibility of doing what Boole proposed, as finishing just ahead of Gregory in the 1837 Tripos math exam, as fourth wrangler, had been the 43-year-old George Green (1793–1841), the poor son of a baker and, like Boole, self-educated.[4] The most significant obstacle for Boole would be the cost, with Gregory informing Boole that, at minimum, expenses at Trinity would be 200 pounds per year (or, more likely, 250 pounds per year). To get an idea of what that meant in 1840, 200 pounds would have paid the salary of the governor of the Bank of England for six months. To attend Cambridge, Boole would also have to close his school and, as he was the sole support of his entire family, that was simply impossible. All thought of enrolling at Cambridge vanished for good.

Writing and publishing continued unabated, however, with Boole's papers growing in both mathematical depth and length. Increasing depth is always good in mathematical papers, of course, but from a practical point of view that's not the case with length. There is a limited amount of space available in a journal, and Boole's lengthening submissions threatened to take it all. Finally, in June 1843, Gregory wrote Boole about his latest effort to tell him that it was simply too elaborate for the *Journal*. But it wasn't all bad news, because Gregory also thought the paper was good enough for submission to the *Transactions of the Royal Society of London*. Now *that* would be the big leagues for Boole!

Gregory's enthusiasm for this latest Boole paper ("On a General Method in Analysis") was at least partly due to the particular topic—the use of symbolic algebra to solve both differential and difference equations (you'll see a difference equation, solved with ordinary high school algebra, in Chapter 7). Gregory had long had intense interest in such an approach, and had himself already published a paper about it. I won't discuss Boole's symbolic algebra for equation solving in this book, as it's his *logical* algebra that will be central for us, but the publication of this latest paper is important because its appearance would change his life. It was with symbolic algebra that Boole began what would develop into the centerpiece of his mathematical work, that of treating *operations* (such as differentiation and differencing) as *symbolic operators* ($\frac{d}{dt}$ and Δ, respectively) that could be manipulated as if they were numbers. This resulted in the operators becoming detached from their arguments and, in fact, symbolic algebra often went under the alternative and descriptive name of "separation of symbols."

After some controversy at the Royal Society on whether to accept the paper or not, it was published—and in November 1844 it earned Boole a Royal Medal (established with the blessing of Queen Victoria) as the best mathematics paper published in the *Transactions* between June 1841 and June 1844. Gregory, alas, never knew this, his tragically early death (probably from cancer) having occurred some months earlier, in February.[5] Boole didn't forget what he owed, however, as in his paper he inserted a footnote saying of Gregory: "Few in so short a life have done so much for science. The high sense which I entertain of his merits as a mathematician, is mingled with feelings of gratitude for much valuable assistance rendered to me in my earlier essays."

Gregory was gone, but Boole had flown the nest and, with his now impressive publication record and a Royal Medal, was a known and respected author. His writing continued to appear without interruption. In late 1847, for example, his interest in symbolic algebra broadened from the purely mathematical to include logic, with the publication of the pamphlet *Mathematical Analysis of Logic*.[6] It would be, in retrospect, a first draft of his masterpiece *An Investigation of the Laws of Thought* which would appear seven years later. Boole almost immediately began the process of tinkering with *Mathematical Analysis*, and in 1848 published a paper version ("The Calculus of Logic") in the *Cambridge and Dublin Mathematical Journal* (the renamed *Cambridge Mathematical Journal*), with William Thomson in the editor's chair (since 1845) as the permanent replacement to Gregory.

In 1849 all of Boole's hard work achieved for him what, as a teenager teaching dull boys simple math at an obscure boarding school in Doncaster, would have been just an outrageous day-dream fantasy. That year he applied for the position of professor of mathematics at the newly created Queen's College (today's University College) in Cork, Ireland. No matter his lack of a university degree: his outstanding publication record, his impressive reputation among fellow mathematicians, and of course his Royal Medal, trumped everything else. He was appointed and, at age 34, he was now *Professor* Boole. His annual stipend was 250 pounds, plus 2 pounds per student per academic term. To anyone who has read of Victorian society and its rigid class system based on inherited wealth and intertwined family connections (none of which Boole enjoyed), this was an absolutely tremendous achievement. His was an appointment based solely on merit, merit so bright and shining that it was simply impossible to ignore. Boole had at last found a home, and he would spend the all-too-few years left to him in Cork.

But before that, Boole had to move, and this time he was to do it by himself. His father had died the year before, and his mother refused to leave England. After making financial arrangements for her care, he prepared to cross the Irish channel, alone, to his new life. Boole's elevated standing in Lincoln is illustrated by the fact that, just before his departure, the city threw a grand public dinner in his honor, during which he was presented with a silver inkstand and a valuable collection of books. It was a happy send-off for the local boy who had succeeded against all odds, and then Boole, for the last time, moved from Lincoln. The town of Lincoln never forgot him, however, and when Boole died fifteen years later the town installed a beautiful stained glass window in the local cathedral in his memory.

Figure 3.1.1. Boole was in London June and July of 1864, just months before his death. While there, he stopped in at the famous London School of Photography at 174 Regent Street, one of the pioneers in commercial Victorian photography, and had this full-length portrait taken. Photo reproduced by arrangement with the Boole Library, Special Collections and Archives, University College, Cork, Ireland.

After settling in at Cork and starting his teaching duties, Boole's mathematical studies added yet another dimension. In 1767 the Reverend John Michell (1724–1793) had published a paper in the Royal Society's *Philosophical Transactions* titled "An Inquiry into the Probable Magnitude and Parallax of the Fixed Stars, from the quantity of light which they afford to us, and the particular circumstances of their situations."[7] This paper, which contains a number of probability calculations, was brought to Boole's attention—did he perhaps recall, as he read it, what he had seen through his father's telescope as a child? In any case, Boole soon after began to publish on probability ("On the Theory of Probabilities, and in Particular on Mitchell's [sic] Problem of the Distribution of Fixed Stars," *The Philosophical Magazine*, June 1851), and in Chapter 6 I'll tell you more about this new interest. By 1858 his continuing work in probability was of such a high caliber that it won him the Keith Prize (a gold medal and 50 pounds) from the Royal Society of Edinburgh for his paper the previous year in the Society's *Transactions* ("On the Application of the Theory of Probabilities to the Question of the Combination of Testimonies or Judgements").

Boole's personal life also experienced a dramatic alteration with the move to Cork. In 1850 he met Mary Everest, who was visiting the family of her uncle John Ryall, who was vice-president and professor of Greek at Queen's College, and a friend of Boole. Mary the niece of Sir George Everest (the surveyor-general of India, after whom the mountain is named), was at age 18 only a bit more than half his age. At first cautious because of the difference in their ages and social class, by 1855 Boole had overcome both concerns and they married. The union was, by all accounts, a happy one, and it produced five daughters for the Booles between 1856 and 1864.

In 1956 Boole's grandson Geoffrey Taylor, related a touching story about what sort of family man Boole was, a story told to him in a letter from Boole's youngest daughter Ethel Voynich.

> My aunt ...wrote me that an old lady who had known Boole in Cork in her youth told her of the following incident. "One day in June, 1856, she went into the slum alley behind the College to engage a chimney sweep for her flues. As she was walking down the alley, she saw father ahead of her, knocking at one door

> after another. She came past him in time to see him passionately shaking hands with a ragged and barefoot man, and saying, "I had to come and tell you, dear friends: I've got a baby [his first child], and she *is* such a beauty."[8]

To many that sort of episode was the reason why, as one historian put it, "In general Boole's reputation among the poorer people of the neighborhood was that of an innocent who should not be cheated. Among the higher classes he was admired as something of a saint, but thought to be rather odd." As an example of that second reaction, when Mary Everest (not yet Mary Boole) asked a local Cork woman where her children were, she was told that George had taken them for a walk and that, while he was indeed a favorite with children, "He is no favorite of mine ... I don't enjoy his society. I don't care to be with such very good people ... he never shows you that he thinks you wicked, but when you are near anyone so pure and holy, you can't help feeling how shocked he must be at you. He makes me feel very wicked; but I am always at ease when the children are with him; I know they are getting some good." Being good with children, and making mothers feel "wicked" by comparison, doesn't mean that Boole should be thought of as having been a shy, gentle person. He could be quite forceful when he thought it was required. He had numerous public battles in the newspapers, for example, with the chemist Sir Robert Kane (1809–1890), the president of Queen's College, over a variety of issues that, like most issues in academic fights, were of a consequence far less impressive than were the fireworks. Boole was hard for Kane to ignore, however, as Boole was clearly one of the college's stars.

Almost from the start, Boole's academic life in Cork was one of moving from one honor and achievement to the next. In 1852 he received an honorary doctorate from the University of Dublin, in 1854 *An Investigation of the Laws of Thought* was published, in 1857 he was elected a Fellow of the Royal Society of London, and there was the Keith Prize in 1858. In 1859 his *A Treatise on Differential Equations* was published, and he received another honorary doctorate, this time from Oxford. In 1860 another textbook was published, *A Treatise on the Calculus of Finite Differences*.

Boole had found heaven in Cork. He loved his teaching, his family, and he was at the peak of his intellectual powers. His life was perfect. Then, on November 24, 1864, Boole made a fatal mistake in judgment. Walking the two miles from his home to the college to deliver a lecture, he was caught in a sudden rainstorm and was drenched to the skin. Rather than take time to dryout, to warmup, to change into dry clothes, to do *anything* that would delay his class, he instead lectured in soaking wet clothes. The result was first a bad cold, and then pneumonia. Mary, a believer in homeopathic medicine (each human ailment should be treated with something resembling the ailment's cause), reasoned that her husband was coughing his life away because he had gotten wet and so put him to bed between cold, wet sheets. Some writers have suggested that she went even beyond that and, in a scene that conjures up any number of bad television comedies, dowsed him with buckets of cold water.

On December 8, 1864, Boole died, not yet fifty years old. Some days later, on December 17 a brief obituary notice appeared in the London literary magazine *Athenaeum* that, after listing *Laws of Thought* as one of Boole's principal works, condescendingly called it a book "which sought a very limited audience, and we believe, found it." There might actually have been some truth to that at the time, and Boole's book *would* continue to slumber quietly in library stacks and on scholars' bookshelves for decades to come. In 1938, however, an event occurred that changed everything. After that year, Boole's name would shine forever, while it is the *Athenaeum* that has vanished.

3.2 THE ELECTRICAL ENGINEER[9]

It is no exaggeration that Claude Shannon was the Father of the Information Age and his intellectual achievement one of the greatest of the 20th century.

—Notices of the American Mathematical Society

Claude Elwood Shannon was born April 30, 1916, in Petoskey, Michigan, to Claude, Sr. (1862–1934), a business man and probate judge, and Mabel Shannon (1880–1945). He was the second of two children. His sister, Catherine (1910–2008), earned a master's degree in mathematics at the University of Michigan and became a professor of mathematics at the College of North Central Illinois. The first sixteen years of Shannon's youth were spent in the small town (3,000 people) of Gaylord, Michigan, where he graduated from Gaylord High School in 1932. His mother was a language teacher at that school and, for a time, principal. Shannon displayed an early interest in "how things work," building model airplanes, a radio-controlled boat, and a telegraph system linking his house to a friend's a half-mile distant (with the help of a strategically located wire fence). He earned pocket money during high school by delivering telegrams and fixing radios at a local department store. (This was common among many youngsters in the 1920s and 1930s who went on to technical careers; a similar story is told by Richard Feynman in his famous 1985 autobiographical *Surely You're Joking, Mr. Feynman!*)

Following in the footsteps of his sister, he enrolled at the University of Michigan, from which he graduated in 1936 with double bachelor's degrees in mathematics and electrical engineering. It was in a class there that Shannon was introduced to Boole's algebra of logic. You can get an idea of Shannon's growing mathematical abilities during those undergraduate years by reading a little note in the *American Mathematical Monthly* for January 1935 (p. 45). There you'll find his solution to an interesting challenge problem in Euclidean geometry, posed the year before by another reader.

In Shannon's senior year at the University of Michigan he began to wonder about what to do next. It was the middle of the Great Depression and, for a person with Shannon's intellect, graduate school was the obvious path. But *where*? Then he spotted an announcement that the MIT Department of Electrical Engineering was looking for a research assistant to work part-time on Vannevar Bush's famous differential analyzer. Shannon applied, got the job, and so scored a double win. Shannon had, in fact, found an early mentor as every bit as important to him as Gregory had been to Boole.

First, Bush (1890–1974) was a man with immense stature in American science, and Shannon would greatly benefit from having him as an early champion. As a measure of Bush's importance, during the Second World War he served, at the direct request of President Roosevelt, first as chairman of the National Defense Research Committee and then, from May 1941, as the director of the Office of Scientific Research and Development, which oversaw such crucial military developments as radar, the proximity fuze, and, until the Army Corps of Engineers took over in May 1943, the construction of the American atomic bomb (the Manhattan Project).

Second, the electromechanical differential analyzer was at that time the world's most advanced analog computer, able to numerically solve very complicated differential equations, and it would be the spark that ignited Shannon's first technical triumph. A complex circuit of over 100 relays controlled the analyzer, and it was part of Shannon's job to understand the controller and to maintain it. After initially pondering this task, followed by spending the summer of 1937 at Bell Labs (New York City), Shannon had his famous epiphany of marrying Boolean algebra with electrical switching circuits. The final, polished result was his MIT master's thesis, "A Symbolic Analysis of Relay and Switching Circuits," which was published in 1938 in the *Transactions of the American Institute of Electrical Engineers*. That work was of such power that it has been labeled by many, in the decades since, as "the most important master's thesis ever written." So impressed by it was Bush, in any case, that he applied his considerable influence in support, and Shannon received the 1940 Alfred Nobel Prize, awarded each year by the combined engineering societies of the United States to the best engineering paper by an author no older than 30.

In late 1938, with his switching thesis done, Shannon moved from the EE department to the math department at MIT. He did that under the advice of Bush, who was then president of the Carnegie Institution in Washington, D.C. Bush made his suggestion because one of the operations funded under his leadership was the genetics laboratory at Cold Spring Harbor in Long Island, New York. Bush felt that Shannon's symbolic success in switching theory (he called it Shannon's "queer algebra") might have a similar payoff in genetics, and so Shannon spent the summer of 1939 at Cold Spring Harbor, working with the well-known researcher Barbara Stoddard Burks (1902–1943), who was a member of the laboratory's Eugenics Records Office.

Eugenics, the study of inherited human traits, would fall into disrepute once it was learned how the Nazis had used it to justify their monstrous program of genocide. Bush was ahead of the world on that score—he closed the eugenics research program at Cold Spring Harbor in 1940 (an action which might have contributed to a depression that caused Burks, at age 40, to commit suicide by jumping 200 feet to her death from the George Washington Bridge in New York City). That work at Cold Spring Harbor led to Shannon's doctoral dissertation ("An Algebra for Theoretical Genetics"), which was never published until it appeared decades later in Shannon's *Collected Papers*. In his 1987 interview in *Omni*, Shannon described his thesis as follows: "My theory has to do with what happens when you have all the genetic facts. One could calculate, if one wanted to . . . the kind of population you would have after a number of generations."[10] Both Burks and Bush strongly urged Shannon to publish, but he had lost interest in the topic, and, besides, he had other, more urgent matters that demanded his attention.

With his PhD in hand, and after spending the summer of 1940 back at Bell Labs, Shannon used a National Research Council Fellowship for a year's stay at the Institute for Advanced Study in Princeton, New Jersey, where he worked under the great mathematician Hermann Weyl. Also there were such luminaries as John von Neumann and Albert Einstein. He might even have bumped into Richard Feynman, who was working on his PhD in physics at Princeton. Also there with Shannon was his first wife, Norma Levor (born 1920), whom he had married in 1939. Theirs was an intense, passionate, but ultimately doomed brief marriage, and Norma left him in June 1941. With all that going on in his life, it isn't surprising that writing up his doctoral dissertation wasn't high on Shannon's list of things to do. It is quite curious (to me, at least) that Norma's name appears in none of the historical essays written by professional colleagues after Shannon's death (did they not know of her, or was it an attempt to protect Shannon's "image"?), and he himself never mentions her in any of the interviews he gave in later years.

Shannon's failed first marriage is of more than prurient interest because, I believe, it gives some insight into his personality then, and perhaps even more insight into his later curious behavior that has always been dismissed as simply being "charmingly odd." The arrival of the two in Princeton must have caused at least some discussion among those at the Institute; after all, the pretty young wife of the new Fellow—who on at least one occasion poured tea for Einstein—was a mere twenty-year-old who had just finished her junior year at Radcliffe College.

After leaving Shannon in 1941, Norma didn't simply fade into history. She moved to Hollywood and became a screenwriter. In 1942 she married fellow screenwriter Ben Barzman (1911–1989), who wrote the screen treatments for the John Wayne film *Back to Baatan* (1945), Charleton Heston's *El Cid* (1961), and George Peppard's *The Blue Max* (1966). She herself wrote the now classic *The Locket* (1946) with its flashbacks within flashbacks, fled to France for nine years with her husband after both were blacklisted in the 1950s by the House Un-American Activities Committee, and became fast friends with actress Sophia Loren (who had appeared in *El Cid*).

Norma was clearly a woman of depth. So, the obvious question: why did she leave Shannon? In her 2003 book, *The Red and the Blacklist: The Intimate Memoir of a Hollywood Expatriate* (which contains a 1939 photograph, never before published, of Shannon sitting in the cockpit of a small plane), she offers us a clue. In the summer of 1963 she was visiting in Cambridge (Shannon was by then a faculty member at MIT) and the two got together for the first time since 1941. The first words Shannon spoke, she relates, were, "Why did you leave me?" to which she replied, "You were sick and wouldn't get help." That rather obscure exchange was clarified (she wasn't referring to a physical illness) years later, in a 2009 talk that you can watch for yourself on www.youtube.com/watch?v=1gv7ywg1H0Q. There she compares Shannon to John Nash of *A Beautiful Mind* fame, a clear reference to a mental disturbance—a disturbance that she says was so pronounced she decided she would not have any children with Shannon. This comparison with Nash, who was clearly a disturbed person, is almost certainly a gross exaggeration (remember, Norma was a Hollywood screenwriter!), but it does illustrate the enormous tension under which Shannon was operating during those years.

At the end of his year at Princeton, and once again alone, Shannon accepted an offer to return to Bell Labs as a full-time member of the technical staff in the mathematical research group. There he would enjoy an astonishingly creative fifteen years, including the production of his masterpiece—what *Scientific American* called "the *Magna Carta* of the information age"—the 1948 "A Mathematical Theory of Communication." Initially his work at Bell Labs dealt with anti-aircraft fire-control systems, the need for which had grown in importance with the appearance of the 400 mph German pulse-jet V1 "flying robot bomb," the world's first cruise missile. (The German V2 rocket—the world's first ballistic missile—is also often lumped in with the V1 as driving fire-control system development during Shannon's day, but it would be quite difficult to shoot down a V2 today, during its 2,000 mph terminal atmospheric reentry phase, much less with 1940s gun technology!)

Later work at Bell Labs took Shannon into the arcane world of cryptography, during which he met the English mathematician Alan Turing (1912–1954), who was a key player in the supersecret British Ultra program ("Ultra" was the code-name for the intelligence obtained from intercepted messages sent by German Enigma coding machines that the Nazis incorrectly thought unbreakable). Turing's impact and influence on Shannon are discussed further in Chapter 9. In 1945 Shannon wrote a classified ("Confidental," which really isn't very 'secret') report, "A Mathematical Theory of Cryptography" which was declassified in 1949 when it appeared in *The Bell System Technical Journal* under the new title of "Communication Theory of Secrecy Systems." Some historians of science have speculated that it was his work in cryptography that led to Shannon's 1948 masterpiece, "A Mathematical Theory of Communication," of which you'll find a (*very*) brief discussion in Chapter 7. Shannon himself, however, was always quite clear on this, crediting papers published in *The Bell System Technical Journal* in the 1920s by Bell Labs scientists Ralph Hartley (1888–1970) and Harry Nyquist (1890–1976); indeed, he specifically credits both men in his "Mathematical Theory." And even before the war, Shannon wrote a letter, dated February 16, 1939, to Bush on "some of the fundamental properties of general systems for the transmission of intelligence" in which the work of Hartley is mentioned.

"A Mathematical Theory of Communication" stunned the engineering world; it was written with such clarity and freedom from obfuscating mathematics (much to the irritation of some pure mathematicians!—see Chapter 6) that real engineers could actually read and understand it. It was simply a tour de force, simultaneously founding the entirely new research field of information theory, posing and solving some extremely difficult problems, and pointing its readers toward other problems that remained unanswered. Einstein is famous for many sayings, but one that particularly applies here is this: "I have little patience with scientists who take a board of wood, look for its thinnest part, and drill a great number of holes where drilling is easy." With his "Mathematical Theory," Shannon drilled a very big hole in a very thick, very hard piece of wood.

If anything, Shannon's *Magna Carta* was perhaps *too* successful in gathering accolades. As one of his Bell Labs colleagues, E. N Gilbert, later wrote in 1966 (see note 1 in Chapter 7):

> Not all the attention came from communication engineers, however ... physicists were interested in the new interpretation of entropy as information [see Chapter 7 for more on this]. Psychologists found that the new information measure gave a convenient quantitative estimate of the difficulty of certain experimental tasks. Other applications have been to linguistics, music, cryptography, and gambling. The response to Shannon's paper was so great that by 1953 the Institute of Radio Engineers formed a Professional Group on Information Theory with a journal of its own [*IRE Transactions on Information Theory*]. ... Information theory was a glamor science for many years. It was popularly supposed that information theory held the key to progress in remote fields to which in fact it did not apply.

Shannon himself fully appreciated what he had wrought, and in 1956 he authored an editorial plea, "The Bandwagon," in the new *IRE Transactions* for a more restrained application of information theory. It fell mostly on deaf ears and blind eyes, however, and it took a second editorial by someone else to have an impact. Written two years later by a young MIT electrical engineering professor, Peter Elias (1923–2001), the hilarious "Two Famous Papers" mocked both ends of the spectrum. The first (fictional) paper "Information Theory, Photosynthesis and Religion," was a laugh at those who thought information theory was applicable to every imaginable problem. And the second (fictional) paper, "The Optimum Linear Mean Square

Filter for Separating Sinusoidally Modulated Triangular Signals from Randomly Sampled Stationary Gaussian Noise, with Applications to a Problem in Radar," poked fun at those who used information theory as simply an exotic way to solve problems already solved years earlier by more traditional methods. Elias's essay had immediate influence on improving the quality of papers in the *IRE Transactions*, and it didn't hurt his reputation one bit—two years later he was appointed head of MIT's electrical engineering department.

In March 1949 Shannon married his second wife, Mary Elizabeth Moore (born 1922), a mathematician he met at Bell Labs. They had three children, two sons and a daughter.

It was during his years at Bell Labs that the first tales of Shannon's less mathematical interests took hold : the riding of a unicycle through the corridors of Bell Labs while juggling balls is perhaps the best known. Shannon was never shy in displaying his interest in toylike gadgetry, with his maze-solving, relay-controlled robotic mouse (see Figure 3.1.1) being nearly as well known as the unicycle/juggling tales. Not so well known was his construction of what has become known as the "Ultimate Machine." Here is how science fiction writer Arthur C. Clarke melodramatically described it in his 1958 nonfiction book *Voice across the Sea* (p.159) when remembering a visit he made to Bell Labs in Murray Hill, New Jersey:

> I cannot leave Bell Labs without mentioning one more device which I saw there, and which haunts me as it haunts everyone who has ever seen it in action. It is the Ultimate Machine—the End of the Line. Beyond it there is Nothing. It sits on Claude Shannon's desk driving people mad. Nothing could look simpler. It is merely a small wooden casket the size and shape of a cigar box, with a single switch on one face. When you throw the switch, there is an angry, purposeful buzzing. The lid slowly rises, and from beneath it emerges a hand. The hand reaches down, turns the switch off, and retreats into the box. With the finality of a closing coffin, the lid snaps shut, the buzzing ceases, and peace reigns once more. The psychological effect, if you do not know what to expect, is devastating. There is something unspeakably sinister about a machine that does nothing—absolutely nothing—except

> switch itself off. Distinguished scientists and engineers have taken days to get over it. Some have retired to professions which still had a future, such as basket weaving, beekeeping, truffle hunting, or water divining.

Those last two words, I think, we can all agree are just a bit over the top. Still, it *is* the perfect illustration of a comment made by Shannon in his *Omni* interview: "I am always building totally useless gadgets ...just because I think they're fun to make." During his visit at Bell Labs, Clarke was so taken by Shannon's fascination with games played by machines that he inserted a throwaway mention of it in his 1956 short story "The Pacifist," one of his "White Hart" English Pub tales. At the start the narrator tells us that

> "I got to the "White Hart" late that evening, and when I arrived everyone was crowded into the corner under the dartboard. All except Drew, that is; he had not deserted his post, but was sitting behind the bar reading.... He broke off ... long enough to hand me a beer and to tell me what was going on.
>
> "Eric's brought in some kind of games machine—it's beaten everybody so far. Sam's trying his luck with it now."
>
> At that moment a roar of laughter announced that Sam had been no luckier than the rest, and I pushed my way through the crowd to see what was happening. On the table lay a flat metal box the size of a checkerboard, and divided into squares in a similar way. At the corner of each square was a two-way switch and a little neon lamp; the whole affair was plugged into the light socket ... and Eric Rodgers was looking round for a new victim.
>
> "What does the thing do?" I asked.
>
> "It's a modification of naughts and crosses—what the Americans call Tic-Tac-Toe. Shannon showed it to me when I was over at Bell Labs."

Shannon's fascination with games wasn't all goofiness. His maze-running mouse, for example, was actually a clever way of illustrating the ability of a machine to explore a strange environment and learn

Figure 3.2.1. This photograph, taken in 1952, shows Shannon with Theseus, his maze-solving "mouse" built in 1950. The mouse was named in honor of the character from Greek mythology who, after killing the Minotaur in the monster's maze (the Labyrinth), found his way back out because he had unrolled a ball of string behind him on the way in. The mouse was moved through a 5-by-5 square, reconfigurable maze by an electromagnet mounted on wheels positioned beneath the floor of the maze. Electric motors powered the wheels, and the motors in turn were controlled by a relay logic circuit (also beneath the floor). The mouse could "explore" the maze according to a fixed strategy that Shannon built into the relay logic, "learning" where the maze walls were by bumping into them. Eventually, the mouse (that is, the relay logic) learned to run, without bumping any walls, through the entire maze. Photo reproduced by arrangement with the MIT Museum, Cambridge, MA.

from its experiences. Shannon was very much interested in machines that were far from being pointless (unlike the so-called Ultimate Machine), and wrote on the possibility of machines playing and winning against humans in chess, a game that has come almost to define intelligence. Despite his own dismissal of games in *Omni,* he made them the entire point of his talk, "Game Playing Machines" upon accepting a medal from the Franklin Institute in 1955, during which he demonstrated some of his creations to the audience, including the maze mouse.

In 1956 Shannon was invited by MIT to be a visiting professor, and in 1958 he left Bell Labs to become the Donner Professor of Science, with a joint appointment in mathematics and electrical engineering. At MIT he did not teach regularly scheduled classes, but instead ran frequent seminars open to all, and supervised a very small number of graduate theses. His fascination with gadgets never waned at MIT, and he did produce a few more results in information theory. Interestingly, however, and in contradiction to his *Omni* declaration of "no interest in money"—a sentiment he repeated in a 1990 *Scientific American* profile ("I've always pursued my interests without much regard to financial value")—at MIT he did develop a strong interest in making money.

More technically, in what is called *portfolio management,* the sort of activity that is the heart and soul of pension and mutual funds. At first ignored by financial professionals, Shannon's ideas (along with those of his Bell Labs colleague John L. Kelly Jr. and the mathematician Ed Thorpe) have today been enthusiastically embraced. The *Transactions on Information Theory* began publishing papers on portfolio theory in the 1980s, and many PhDs in information theory have since found employment with Wall Street investment firms. Shannon himself became wealthy by applying his ideas to his personal finances, a story you can read at length in the 2005 book by William Poundstone, *Fortune's Formula.*

Shannon retired from MIT in 1978 and withdrew from academic activity. He had received many honors in his life, including a 1966 National Medal of Science presented to him at the White House by President Lyndon Johnson, and he had no ego-need for additional recognition. He rarely attended meetings, and his absence only added to his growing legend as a mysterious genius. One meeting he did attend was the 1985 International Information Symposium held in Brighton, England. At first no one paid any attention to the tall, quiet, slender man who appeared to simply wander in and out of the technical sessions. Word soon got out that Shannon was there, however, and when he was eventually identified, the mob of excited engineers that crowded about Shannon, eager to get his autograph, seemed to one observer to be "as if Newton had showed up at a physics conference."

It was at that meeting, too, that some people began to notice that all was not quite right with Shannon. In the years that followed he had increasing difficulty with his memory, to the point of forgetting where he was going or how to return home when out driving. By 1993 dementia had so impaired his ability to function that he was placed in a Medford, Massachusetts, nursing home. There the years passed and he remained unaware of the almost daily advances in communications and computer technology in which his work was fundamental. He completely missed, for example, the creation of the World-Wide Web, a development he would surely have taken immense pride in helping make possible. By all accounts, from those who visited him, he remained a happy, cheerful man, but one who could not recognize his own handwritten papers. And there, on February 24, 2001, just shy of his eighty-fifth birthday, Claude Shannon died a death perhaps even more cruel than had been Boole's. He had lived his life blessed with a brain of rare magnificence, but in the end Alzheimer's disease had taken it all away.

But his work remains and continues to inspire. When you enter the lobby of Bell Labs in Murray Hill, you pass between two busts in honor of its two greatest men. One, as you might expect, is of Alexander Graham Bell. The other—from an institution that has produced numerous Nobel laureates in physics and so has its pick of giants about which to boast—is of Claude Elwood Shannon.

NOTES AND REFERENCES

1. The first part of Chapter 3 is based on numerous biographical essays and one book about Boole's life and work. Much of all that secondary material is repetitive, and not always in agreement, and to cite each source as it appears in the text would lead to an explosion of notes. So I'll limit myself with simply listing once, here, all of the sources that I found most useful in my writing:

(a) Geoffrey Taylor, "George Boole (1815–1864)," *Notes and Records of the Royal Society of London*, August 1956, pp. 44–52. Taylor was Boole's grandson.

(b) Elizabeth B. Cooksey, "George Boole: The Man Behind And/Or/Not," *Libraries & Culture*, Winter 1997, pp. 81–93.

(c) William Kneale, "Boole and the Revival of Logic," *Mind* , April 1948, pp. 149–175.

(d) T.A.A. Broadbent, "George Boole (1815-1864)," *Mathematical Gazette*, December 1964, pp. 373–378.

(e) Rush Rhees, "George Boole As Student and Teacher by Some of His Friends and Pupils," *Proceedings of the Royal Irish Academy* 57, 1954–1956, pp. 74–78.

(f) R. H., "George Boole FRS," *British Quarterly Review*, July 2, 1866, pp. 141–181. "R. H." was Boole's friend, the Reverend Robert Harley (1828–1910).

(g) Desmond MacHale, *George Boole, His Life and Work*, Boole Press, Dublin, Ireland, 1985.

(h) George Boole, *Studies in Logic and Probability* (edited by Rush Reeves), Watts & Co., 1952.

2. The calculus of variations is both beyond the scope of this book and irrelevant to our concerns here on Boolean algebra but, if you're interested, you can find an extended discussion of both its theory and its classic problems dating back to ancient times before Christ, in my book *When Least Is Best*, Princeton University Press, 2004; corrected paperback, 2007, pp. 200–278. The calculus of variations paper was not Boole's first scientific publication; his private study of Newton's *Principia* in the Mechanics Institute reading room had greatly impressed the Lincoln locals. Enough, in fact, that he was asked to give an address (on February 5, 1835, only a few months past his nineteenth birthday) on Newton during the presentation ceremony of a bust of Newton to the Institute, and that address was soon after published as a pamphlet, *On the Genius and Discoveries of Sir Isaac Newton*, now quite rare.

3. An amazing example of what Gregory's *Journal* would publish that almost certainly would not have been accepted in a more established publication was the first paper ever written by William Thomson (1824–1907), later Lord Kelvin. In 1837 the Anglican cleric and mathematician Philip Kelland (1808–1879) published the book *Theory of Heat*, in which he took sharp exception to central results in Fourier's 1822 theory of the possibility of expanding periodic functions in terms of sinusoids (a routine mathematical tool today: see my *Dr. Euler's Fabulous Formula*, Princeton University Press, 2006; corrected paperback, 2011, pp. 114–187). Thomson, then just a teenager, had read Fourier ("In a fortnight I had mastered it—gone right through it") and

thought it brilliant (he characterized Fourier's work as a "mathematical poem"). He therefore wrote a sharp rebuttal to Kelland's position, and Gregory published it in 1841. Kelland backed off his assertions, and his reputation as a mathematician suffered.

4. Green's name is famous in mathematical physics, and is attached to such things as *Green's functions* and *Green's theorem*. You can find more on his somewhat mysterious life in D. M. Cannell, *George Green, Mathematician and Physicist, 1793–1841: The Background to His Life and Work*, SIAM, 2001.

5. A very nice discussion of symbolic algebra, and of the relationship between Boole and Gregory, is in Patricia R. Allaire and Robert E. Bradley, "Symbolical Algebra as a Foundation for Calculus: D. F. Gregory's Contribution," *Historia Mathematica*, November 2002, pp. 395–426. Another very nice discussion, both technically and historically, of Boole's symbolic manipulation of the differentiation operator $D_x(=\frac{d}{dx})$ is in Michael A. B. Deakin, "Boole's Mathematical Blindness," *Mathematical Gazette*, November 1996, pp. 511–518. You can find even more on D-operators in *Dr. Euler's Fabulous Formula* (see note 3), pp. 92–113, where I use them to prove the irrationality of π^2 (which also proves the irrationality of π because if π were rational then π^2 would be, too—which it isn't). You can find more on the difference operator Δ in my paper "A Simple Operator Approach for Analytic Solution of Linear Constant-Coefficient Difference Equations," *IEEE Transactions on Education*, December 1968, pp. 234–239.

6. The cause of this new interest was a war of words, including a nasty charge of plagiarism, between the English mathematician Augustus De Morgan (1806–1871) and Scottish philosopher William Hamilton (1788–1856), not to be confused with the Irish mathematician of the same name. Hamilton the philosopher thought little of mathematics ("mathematics freeze and parch the mind"), and when De Morgan got caught up in a battle with him over a question in logic, Hamilton's verbal knives came out. Watching on the sidelines at first, Boole quickly concluded that De Morgan was right and the pamphlet was Boole's contribution. We'll hear from De Morgan in the next chapter.

7. Michell's name is famous among physicists today for his 1783 speculation on the existence of stars so massive that their light could not escape the pull of their gravity. That is, almost a century before the birth of Einstein the concept of a black hole was born. For more on Michell, see my *Mrs. Perkins's Electric Quilt*, Princeton University Press, 2009, pp. 139, 165–166.

8. Boole's youngest daughter Ethel Voynich was certainly *not* her father's child in personality. In 1895, after having married a Polish revolutionary, she had (at least according to some writers) a romantic fling with the notorious British superspy Sidney Reilly, who was executed in 1925 by the Russian secret police and is commonly thought to be the model for Ian Fleming's fictional James Bond. That affair was, at least according to some writers, the

inspiration for her 1897 novel *The Gadfly*, which sold millions of copies in Russia and China and was made into a 1955 Russian movie. The musical score for that movie, written by Shostakovich, was *The Gadfly Suite* and its "Romance" movement in particular was the theme music for the 1983 BBC television miniseries *Reilly, Ace of Spies*. The last of Boole's immediate family, Ethel Voynich, died—after an undeniably *very* adventurous life—in 1960 in New York City at age 96.

9. There is, as I write, no book-length biography of Shannon. The second part of Chapter 3 is, therefore, based only on several short biographical remembrances, interviews Shannon gave some years before his death, and obituary notices. The most useful to me were

(a) Robert G. Gallager, "Claude Elwood Shannon," *Proceedings of the American Philosophical Society*, June 2003, pp. 187–191.

(b) Anthony Liversidge, "Profile of Claude Shannon," *Omni* , August 1987 (reprinted in Shannon's *Collected Papers*, N.J.A. Sloane and Aaron D. Wyner, editors, IEEE Press, 1993).

(c) Solomon W. Golomb et al., "Claude Elwood Shannon (1916–2001)," *Notices of the American Mathematical Society*, January 2002, pp. 8–16.

(d) James F. Crow, "Shannon's Brief Foray into Genetics," *Genetics*, November 2001, pp. 915–917.

10. Liversidge, "Profile of Claude Shannon."

4

Boolean Algebra

0101100101100101100010

They who are acquainted with the present state of the theory of Symbolical Algebra, are aware that the validity of the processes of analysis does not depend upon the interpretation of the symbols which are employed, but solely upon the laws of their combination.

—George Boole, in the opening to his *Mathematical Analysis of Logic* (1847)

4.1 BOOLE'S EARLY INTEREST IN SYMBOLIC ANALYSIS

As the above quotation shows, Boole was interested in symbolic analysis years before he wrote his *Laws of Thought*. As mentioned in the previous chapter, before even the *Mathematical Analysis of Logic* had appeared he had published a paper on how to apply the symbolic manipulation of the differentiation and difference operators to the solution of differential and difference equations. The solution in that way of such equations, while of immense importance in mathematical physics, is both outside the scope and beyond the technical level of this book and I won't pursue that mathematics here. My point is simply that the symbolic methods of Boole didn't leap full-born from his head with the *Laws of Thought*. Indeed, others before him—in particular, the German mathematician Gottfried Wilhelm Leibniz (1646–1716)—had pursued a similar goal of reducing logic to algebra, but it was Boole who finally succeeded. What Boole described in his books is *not* exactly what modern users call Boolean algebra, but nevertheless it is from Boole that the modern presentation springs.[1]

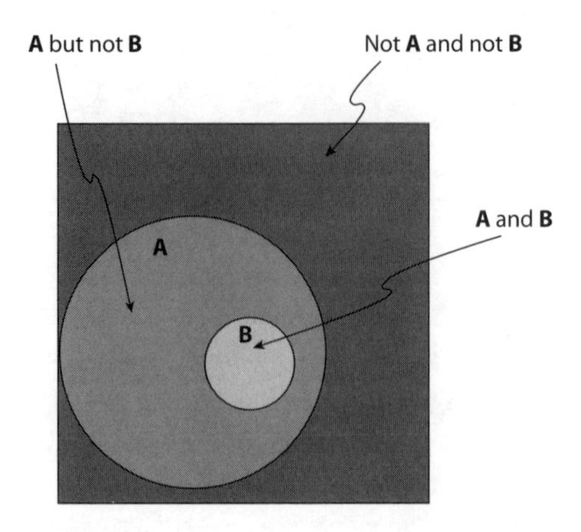

Figure 4.2.1. Every element of **B** is an element of **A**.

4.2 VISUALIZING SETS

To understand the essence of what Boole did to cast logic into algebraic form, it is very helpful to use the language of sets. I'll discuss what Boole did in the next section but, first, to lay the groundwork for that I'll say just a bit here on set terminology. A *set* is simply a collection of things (the *elements* of the set), either physical or conceptual. For example, we could talk of the set of all rocks (I'll call that set **A**) and we could talk of the set of all small rocks (I'll call that set **B**). Following the example of the Swiss-born mathematician Leonhard Euler (1707–1783), we could then pictorially represent **A** and **B** as shown in Figure 4.2.1 (Euler was drawing such diagrams before1770). Every element of **B** is also an element of **A** because every small rock is a rock, but the converse is not true (there are elements of **A** that are not in **B**); all rocks are not small rocks. Notice, too, that there is a region in the diagram that contains those elements that are not in either **A** or in **B**; the elements in that region are all the things that are *not* rocks. Since **B** is completely contained in **A**, there are just three distinct regions in Figure 4.2.1. That is not the most general case, however: if **A** and **B** are sets that partially overlap (intersect), then there is a fourth region, as shown in Figure 4.2.2.

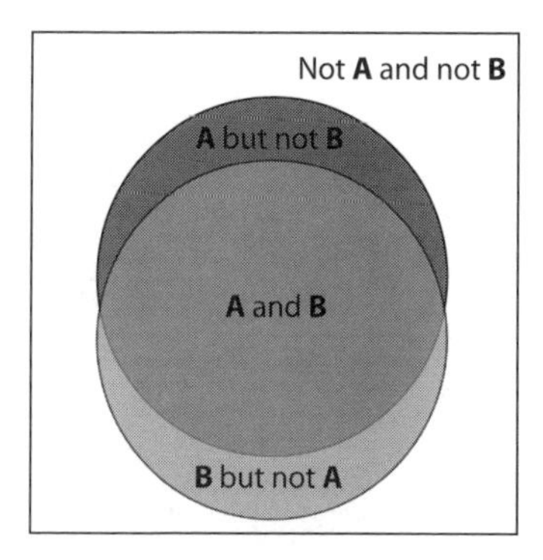

Figure 4.2.2. The four general regions of two sets.

4.3 BOOLE'S ALGEBRA OF SETS

To see how we can achieve Boole's aim of arriving at the laws that "combine" symbols in a logically consistent manner, independent of what those symbols may actually represent, we'll use some intuitive visualizations of sets that I think you'll find easy to accept. Specifically, as one modern analyst wrote,

> From a "universal" or fundamental set U, say the inhabitants of London, we can symbolise sub-sets by means of selective operators. Thus $x(U)$ may denote the set of blue-eyed inhabitants of London, and $y(U)$ the set of left-handed inhabitants of London, so that $x()$ and $y()$ are selective operators. Then $x(y(U))$ will select the blue-eyed left-handers, and $y(x(U))$ the left-handed blue-eyed people. Since these two sets are composed of exactly the same elements, we can write $x(y(U)) = y(x(U))$, or, since the "universe" U is understood, this can be abbreviated to $xy = yx$. If a third operator z, say, is defined, as for instance, selecting all males, we can show that the operations are associative, $x(yz) = (xy)z$.[2]

Notice, carefully, that while these examples have given specific interpretations to the "universe," and to the operators working in that universe, we have already arrived at one of the *general* combination laws for those operators. The same author then continues:

> It can also be observed that this symbolic product [that is, $xy = yx$] plays the part of the logical connective "and", since xy denotes those who are left-handed *and* blue-eyed. But already a difference from the numerical product can be seen; the symbol $x(x(U))$ denotes the selection of all the blue-eyed, followed by the selection of all the blue-eyed, which merely selects all the blue-eyed, or, in symbols, $x(x(U)) = x(U)$, and, briefly, $xx = x$.

Now *this* law is a bit unsettling! In high-school algebra $x^2 = x$ is perfectly fine, but what could it possibly mean in logic? Boole himself thought $x^2 = x$ to be a central result of his symbolic logic. Here's one way, admittedly a bit off-the-wall, to arrive at a possible interpretation. Write $xx = x$ as $x - xx = 0$ or, $x(1 - x) = 0$. Then, rewrite $1 - x$ as $U - x(U)$, which we already know means "all the inhabitants of London, *less* all the blue-eyed inhabitants of London," which, in turn, means "all the inhabitants of London who are *not* blue-eyed." Thus, $x(1 - x) = 0$ means the set "all the people who are blueeyed *and* not blueeyed" is the *empty* set, where we interpret 1 to mean the set of all people in London (the "universe") and 0 to mean the so-called *null* set, which has no elements in it. Since being blueeyed and not being blueeyed are mutually exclusive properties, we shouldn't be surprised to have arrived at a set with no elements in it.

In logic, if x is a statement, then we will take $x = 1$ to mean the statement is true and $x = 0$ to mean the statement is false. And this makes sense on *two* levels: (1) $x^2 = x$ means "true *and* true gives true" (with $x = 1$), and "false *and* false gives false" (with $x = 0$), and (2) in high-school algebra $x^2 = x$ is valid for just two particular numerical values of x, $x = 0$ and $x = 1$. Be sure to note that we are using the symbols 0 and 1 in two quite different ways: (1) as the so-called *logical* 0 and 1 for *false* and *true,* respectively, and (2) as the usual *numerical* 0 and 1 from arithmetic. The fact that (0)(0) = 0 and (1)(1) = 1 with either interpretation is a fortuitous "accident." We'll have yet a third interpretation for 0 and 1, Shannon's electrical one, in the next chapter.

The use of the product xy to represent "x and y" is the notation of electrical engineers (and many mathematicians), and it is the notation I am going to continue to use in this book. Philosophers (and many mathematicians) typically use the symbol ∩ to denote what is called in set theory the *intersection* of sets **A** and **B**. That is, **A**∩**B** are those elements that are in **A** *and* in **B** (the overlap region of **A** and **B** in Figure 4.2.2). So, they would write $x \cap y$ to denote 'x *and* y.' I personally don't like the ∩ notation and, since I am the author, I get to choose! But, just to say it here *once*,

$$xy = x \cap y. \tag{4.3.1}$$

Another logical connective of great use is *or*. There are, in fact, two versions of this connective. I'll use the arithmetic addition sign to denote the so-called *inclusive-or*. For example, I'll write $x + y$ to mean "all the blue-eyed people or all left-handed people or all blue-eyed and left-handed people." The *exclusive-or*, on the other hand, will be written as $x \oplus y$, which means "all the blue-eyed people or all left-handed people but not people who are both blue-eyed and left-handed." (Boole, however, used our inclusive-or + symbol for his exclusive-or!) Philosophers use the ∪ symbol (which I don't like either) to write the set operation of *union* (the entire shaded region of Figure 4.2.2, which is the inclusive-or), and so, just to say it here once,

$$x + y = x \cup y. \tag{4.3.2}$$

If we view 1 and 0 as representing statements that are *true* and *false*, respectively—the view that will be central to the rest of this book—then we can write the following for the *and* connective (where I've used a dot (·)—which in fact I'll usually *not* include when we use letters for so-called *Boolean variables* in the next sections, to emphasize that we are talking about a logical *product*):

$$\begin{aligned} 0 \cdot 0 &= 0 \\ 0 \cdot 1 &= 0 \\ 1 \cdot 0 &= 0 \\ 1 \cdot 1 &= 1, \end{aligned} \tag{4.3.3}$$

which looks, as observed before, just like ordinary arithmetic (but it isn't!). We get just a bit of a shock with the inclusive-or, however:

$$\begin{aligned} 0+0&=0\\ 0+1&=1\\ 1+0&=1\\ 1+1&=1. \end{aligned} \tag{4.3.4}$$

The last line of (4.3.4) makes it very clear that we are *not* dealing with ordinary arithmetic. And finally, for the exclusive-or, we have

$$\begin{aligned} 0\oplus 0&=0\\ 0\oplus 1&=1\\ 1\oplus 0&=1\\ 1\oplus 1&=0. \end{aligned} \tag{4.3.5}$$

4.4 PROPOSITIONAL CALCULUS

The use of the historical phrase *propositional calculus* may seem quite impressive, but the actual mathematics involved is simply that of the previous section, that is, arithmetic (but not ordinary arithmetic). We will use the symbols 1 and 0 to denote statements (propositions) that are *true* and *false*, respectively, obeying the combination laws of (4.3.3), (4.3.4), and (4.3.5). I'll use capital letters to denote statements that are either true or false, e.g., $A=1$ and $B=0$ mean statement A is true and statement B is false. And finally, to complete our development of the mathematics of modern Boolean algebra, there is one last logical operation I need to introduce: the *negation* or *not* operation. It is pretty simple. Whatever the condition of statement A may be (either true or false), not-A (written as $\overline{A}$) is the opposite or *complement*. So if $A=1$ then $\bar{A}=0$, and if $A=0$ then $\bar{A}=1$.

Now, perhaps to your surprise, *that's it*. There is nothing more to Boolean algebra than what I've already told you. All further embellishments —of which there are just a few more to come—follow from what you have already seen. For example, here are some useful Boolean algebraic identities that may, at first glance, look exotic, but in fact you should be able to see their validity either by inspection or after, perhaps, just a bit of pen-and-paper jotting. If A and B are any two statements that are either true or false, then

$$A\bar{A} = 0 \tag{4.4.1}$$

$$AA = A, \tag{4.4.2}$$

$$A + \bar{A} = 1, \tag{4.4.3}$$

$$A + 1 = 1, \tag{4.4.4}$$

$$A + 0 = A, \tag{4.4.5}$$

$$A + A = A, \tag{4.4.6}$$

$$A + AB = A, \tag{4.4.7}$$

$$\overline{AB} = \bar{A} + \bar{B}, \tag{4.4.8}$$

$$\overline{A + B} = \bar{A}\bar{B}. \tag{4.4.9}$$

All of these identities are easy to prove because in Boolean algebra, unlike in the usual high school algebra where variables can potentially have *any* value, Boolean variables can have just one of two values (0 or 1). So, to confirm an identity, all we have to do is examine, one by one, all the possible combinations of variable values (a finite number). In (4.4.1), for example, we have $1 \cdot 0 = 0$ if $A = 1$, or $0 \cdot 1 = 0$ if $A = 0$, and so we see that the right-hand side is 0 independent of A. In (4.4.6), for another example, we have $0 + 0 = 0$ if $A = 0$, and $1 + 1 = 1$ if $A = 1$, and so we see that the right-hand side is the same value as the value A happens to have. The case of (4.4.7) is just a bit more involved because now we have two Boolean variables. The method of proof is unchanged, however, as again we simply evaluate the left-hand side and then the right-hand side of the claimed identity for all possible values of A and B, to get

A	B	$A + AB$
0	0	0
0	1	0
1	0	1
1	1	1

and we see that the column for $A+AB$ does indeed match the column for A. Of course, we could also easily prove this identity by simply writing $A+AB=A(1+B)$ or (because $1+B=1$) $= A\cdot 1=A$.

This sort of table is called a *truth table*, and it is a common tool for proving identities that involve multiple variables. Since Boolean variables are *binary*-valued variables, there are 2^n rows in such a table if we have n variables. For $n=4$ variables, for example, there are only 16 rows, and so evaluating the left-and right-hand sides of any claimed identity is not really a burdensome task. The last two identities, (4.4.8) and (4.4.9), are particularly nice examples of this technique:

A	B	$\overline{AB}$	$\bar{A}+\bar{B}$	$\overline{A+B}$	$\bar{A}\bar{B}$
0	0	1	1	1	1
0	1	1	1	0	0
1	0	1	1	0	0
1	1	0	0	0	0

and we see that the columns for $\overline{AB}$ and $\bar{A}+\bar{B}$ match, as do the columns for $\overline{A+B}$ and $\bar{A}\bar{B}$. These two identities are useful in the design of practical electronic logic circuits (something we'll do in Chapter 7); they are *not* due to Boole but rather to another English mathematician, Augustus De Morgan (Professor of Mathematics at University College, London), who formulated them in 1858.[3] They are known today as *De Morgan's theorems*.

In the truth table I constructed to show that $A+AB=A$, the first two columns listed all possible values for the Boolean variables A and B, and the third column had the values for the *combinatorial Boolean function* $F(A, B) = A + AB$, where F itself is a Boolean variable, too. Given the two variables A and B, this particular function is just one of numerous possible functions connecting A and B. So, here is a natural question to ask: how many functions of two variables are there? There are 4 rows in the truth table for two input variables A and B, and for each row the output function value F could be either a 0 or a 1. So, there are $2^4 = 16$ combinatorial functions of two variables.

Some of these 16 functions are more interesting than others. The two functions $F = 0$ independent of A and B, and $F = 1$ independent of A and B, are (I think) pretty obviously *not* very interesting! One function that *is* interesting is $F = 1$ when $A = B = 0$ and $F = 0$ otherwise. That is, $F = \bar{A}\bar{B}$, which is sometimes called by its old-fashioned name —the *Pierce function*—after the American logician and philosopher Charles Pierce (1839–1914). From De Morgan's theorem we see that it is $\overline{A + B}$. The Pierce function is known today as the *not-or*, or NOR.

Another interesting function is $F = 0$ when $A = B = 1$ and $F = 1$ otherwise. That is, $F = \bar{A} + \bar{B}$, which is sometimes called by its old-fashioned name—the *Sheffer stroke function*—after the American logician and philosopher Henry Sheffer (1882–1964). (The word *stroke* was used because philosophers once wrote it—and maybe some still do!—with a stroke symbol, that is, as A/B. I think this notation, however, is pretty much obsolete today.) From De Morgan's theorem we see that it is $\overline{AB}$. The Sheffer function is known today as the *not-and*, or NAND. We'll return to both the NOR and the NAND functions in Chapter 5, where you'll see they have particularly important places in modern digital circuitry.

More generally, there are 2^{2^n} functions of n variables. This is an expression that gets very big, very fast:

Number of Variables	Number of Functions
1	4
2	16
3	256
4	65,536
5	4,294,967,296
6	$1.84 \cdot 10^{19}$

To gain an appreciation for the last line, if one could build six-variable Boolean functions at the rate of one billion per second, then it would take more than 583 years to build them all.

4.5 SOME EXAMPLES OF BOOLEAN ANALYSIS

I think you are now ready to see how the mathematical machinery we have developed can be put to good use. So, let's use it to solve the four puzzles I gave you in the introduction, starting with Puzzle 1. I'll ask you to look back there to refresh your memory of the details for each puzzle.

PUZZLE 1 SOLUTION

Define the nine Boolean variables X_y, where $X = A, B$, or C and $y =$ red, blue, or white, such that $X_y = 1$ means the statement "box X contains the chip with color y" is true, and $X_y = 0$ means that statement is false. For example, $A_r = 1$ means the statement "box A contains the red chip" is true, while $A_r = 0$ means the statement is false. There are numerous equations that we could now write using this notation, but not all will be useful. One, in particular, however, is

$$A_r + B_r + C_r = 1, \tag{4.5.1}$$

which states the obvious: the red chip *must* be in one of the boxes. The three statements (of which only one is true) given in the introduction are, symbolically, denoted by the variables A_r, $\bar{B}_r$, and $\bar{C}_b$. Now, using the same "careful reasoning" used in the solution given in the introduction, that is, let's take each of the three statements, in turn, as the true one and then reverse the other two, and we arrive at

$$A_r B_r C_b + \bar{A}_r \bar{B}_r C_b + \bar{A}_r B_r \bar{C}_b = 1 \tag{4.5.2}$$

since one of the three logical products will be 1. Obviously $A_r B_r = 0$ since the red chip can't be in both box A and box B, and so the first term in (4.5.2) is 0, giving

$$\bar{A}_r \bar{B}_r C_b + \bar{A}_r B_r \bar{C}_b = 1. \tag{4.5.3}$$

From the logical product of (4.5.1) and (4.5.3) we get

$$(A_r + B_r + C_r)(\bar{A}_r \bar{B}_r C_b + \bar{A}_r B_r \bar{C}_b) = 1$$

or, expanding,

$$A_r \bar{A}_r \bar{B}_r C_b + A_r \bar{A}_r B_r \bar{C}_b + B_r \bar{A}_r \bar{B}_r C_b + B_r \bar{A}_r B_r \bar{C}_b + C_r \bar{A}_r \bar{B}_r C_b + C_r \bar{A}_r B_r \bar{C}_b = 1. \quad (4.5.4)$$

Since $A_r \bar{A}_r = B_r \bar{B}_r = 0$, and since $C_r C_b = C_r B_r = 0$ (these conditions should all now be obvious!), we are left with just one non-zero term (the fourth one) on the left-hand side of (4.5.4):

$$\bar{A}_r B_r \bar{C}_b = 1 \quad (4.5.5)$$

because $B_r B_r = B_r$. Now, for a logical product to be 1, each individual factor must be 1, and so (4.5.5) immediately tells us that $\bar{A}_r = 1$ $(A_r = 0)$, $B_r = 1$, and $\bar{C}_b = 1 (C_b = 0)$. That is, box A does *not* have the red chip, box B does have the red chip, and box C does *not* have the blue chip. So, box C must have the white chip, leaving the blue chip for box A.

PUZZLE 2 SOLUTION

Define the six Boolean variables H, J, D, G, C, and T as follows: if $H = 1$ then the statement "Harry did it" is true, while if $H = 0$ that statement is false. And so on, in a similar way, for the other five variables. Since for one of the pairs of statements *both* statements are false, we have

$$(C + G)(D + T)(T + C)(H + C)(D + J) = 0 \quad (4.5.6)$$

because, while four of the factors must be 1 (those factors associated with the pairs of statements in which one of statements is true), the remaining factor—the one associated with the pair of statements that are both false—must be 0. We, of course, have no idea (yet) which factor is the "remaining factor." If we start to expand (4.5.6), we have

$$(CD + GD + CT + GT)(TH + CH + TC + C)(D + J) = 0. \quad (4.5.7)$$

Now, before continuing, here's a crucial observation: since there are only *two* boys who "did it," then any logical product of three or more different variables will necessarily be 0. This allows us to immediately write as we expand (4.5.7), by inspection,

$$CDC(D+J)=CDCD+CDCJ=CD+CDJ=CD(1+J)=0$$

and so

$$CD=0. \tag{4.5.8}$$

Next, as mentioned earlier we know that four of the five factors in (4.5.6) are 1 and so, if we form the five possible logical products taking four factors at a time, the logical sum of these products will be 1. Thus,

$$\begin{aligned}
&(C+G)(D+T)(T+C)(H+C)+(C+G)(D+T)(T+C)(D+J)\\
&+(C+G)(D+T)(H+C)(D+J)+(C+G)(T+C)(H+C)(D+J)\\
&+(D+T)(T+C)(H+C)(D+J)=1.
\end{aligned} \tag{4.5.9}$$

If we begin to multiply-out the left-hand side of (4.5.9), we have

$$\begin{aligned}
&(CD+DG+CT+GT)(TH+CH+TC+C)\\
&+(CD+GD+CT+GT)(TD+CD+TJ+CJ)\\
&+(CD+GD+TC+GT)(HD+CD+HJ+CJ)\\
&+(CT+GT+C+GC)(HD+CD+HJ+CJ)\\
&+(DT+T+CD \mid TC)(HD+CD+IIJ+CJ)=1.
\end{aligned}$$

Using once again the observation that the logical product of more than two different variables is always 0 *in this problem*, as well as our earlier result of (4.5.8)—that $CD=0$—we can now immediately write, by inspection,

See author's note

and so $C = 1$ and $J = 1$, and we have identified our two thieves as Charlie and James. Without Boolean algebra, I think this problem would reduce the brains of most people to mush!

PUZZLE 3 SOLUTION

Define the sixteen Boolean variables X_i, where $X = A$ (for Alice), B (for Brenda), C (for Cissie), or D (for Doreen), and $i = 1, 2, 3, or 4$, which means X was in i-*th* place in the scholarship competition. For example, $A_1 = 1$ means the statement "Alice was first" is true, while $B_3 = 0$ means the statement "Brenda was third" is false. From Alice's pair of statements we have

$$C_1B_2 = 0 \tag{4.5.10}$$

because one of her two statements is false. But, because one of her statements is true we also have

$$C_1 + B_2 = 1. \tag{4.5.11}$$

Using De Morgan's theorem on (4.5.10) gives

$$\overline{C_1B_2} = \bar{0} = 1 = \bar{C}_1 + \bar{B}_2,$$

and if we logically multiply this result with (4.5.11) we get

$$(\bar{C}_1 + \bar{B}_2)(C_1 + B_2) = 1 \cdot 1 = 1 = \bar{C}_1C_1 + \bar{B}_2C_1 + B_2\bar{C}_1 + \bar{B}_2B_2,$$

or, since $\bar{C}_1C_1 = \bar{B}_2B_2 = 0$, then

$$\bar{B}_2C_1 + B_2\bar{C}_1 = 1. \tag{4.5.12}$$

In a similar fashion, from Brenda's statements we have

$$C_2D_3 = 0, \quad C_2 + D_3 = 1,$$

from which it follows that

$$\bar{C}_2D_3 + C_2\bar{D}_3 = 1. \tag{4.5.13}$$

And from Cissie's statements we have

$$D_4A_2 = 0, \quad D_4 + A_2 = 1,$$

from which it follows that

$$\bar{A}_2D_4 + A_2\bar{D}_4 = 1. \tag{4.5.14}$$

Logically multiplying (4.5.12), (4.5.13), and (4.5.14) together, we have

$$(\bar{B}_2C_1 + B_2\bar{C}_1)(\bar{C}_2D_3 + C_2\bar{D}_3)(\bar{A}_2D_4 + A_2\bar{D}_4) =$$
$$(\bar{B}_2C_1\bar{C}_2D_3 + B_2\bar{C}_1\bar{C}_2D_3 + \bar{B}_2C_1C_2\bar{D}_3 + B_2\bar{C}_1C_2\bar{D}_3)$$
$$(\bar{A}_2D_4 + A_2\bar{D}_4) = 1.$$

Since $C_1C_2 = B_2C_2 = 0$—Cissie couldn't have been both first *and* second, and Brenda and Cissie can't *both* be second—we have the reduction to

$$(\bar{B}_2C_1\bar{C}_2D_3 + B_2\bar{C}_1\bar{C}_2D_3)(\bar{A}_2D_4 + A_2\bar{D}_4) = 1. \tag{4.5.15}$$

Continuing with the expansion of (4.5.15),

$$\bar{A}_2D_4\bar{B}_2C_1\bar{C}_2D_3 + \bar{A}_2D_4B_2\bar{C}_1\bar{C}_2D_3 + A_2\bar{D}_4\bar{B}_2C_1\bar{C}_2D_3$$
$$+A_2\bar{D}_4B_2\bar{C}_1\bar{C}_2D_3 = 1.$$

Now, the first and second terms on the left are 0 (Doreen can't be both third *and* fourth), and the fourth term is 0 (Alice and Brenda can't *both* be second). That is, all but the third term are each 0, and so

$$A_2\bar{D}_4\bar{B}_2C_1\bar{C}_2D_3 = 1. \tag{4.5.16}$$

Each and every factor in the logical product of (4.5.16) must be 1, and that says $A_2 = 1$ (Alice was second), $C_1 = 1$ (Cissie was first, and this is consistent with $\bar{C}_2 = 1$; Cissie was *not* second), $D_3 = 1$ (Doreen was third, and this is consistent with $\bar{D}_4 = 1$; Doreen was *not* fourth), and so we are left with fourth place for Brenda, which is consistent with

$\bar{B}_2 = 1$ (Brenda was *not* second). Clearly, this is yet another problem where "your brain would be mush" without Boolean algebra.

PUZZLE 4 SOLUTION

This is actually the easiest of the four puzzles to solve, *if* you have the idea of using a truth-tablelike tabulation of all possible cases for four Boolean variables. That is, writing $A = 0$ to mean "A doesn't hunt" and $C = 1$ to mean "C hunts," and similarly for B and D, we can write the following table of 16 rows listing all possible combinations of hunting/no hunting for each of the four hunters on any given day. Row 6, for example, says that B and D hunt, while A and C don't.

Row	A	B	C	D
1	0	0	0	0
2	0	0	0	1
3	0	0	1	0
4	0	0	1	1
5	0	1	0	0
6	0	1	0	1
7	0	1	1	0
8	0	1	1	1
9	1	0	0	0
10	1	0	0	1
11	1	0	1	0
12	1	0	1	1
13	1	1	0	0
14	1	1	0	1
15	1	1	1	0
16	1	1	1	1

Now, condition (1) says that if $A = 1$ then $B = 0$. That eliminates rows 13, 14, 15, and 16. Condition (2) says that if $B = 1$ then $D = 1$ and $C = 0$. That further eliminates rows 5, 7, and 8. And condition (3) says that if $D = 1$ then $A = 1$ or $B = 1$ (or *both* A and $B = 1$). That further eliminates rows 2 and 4. We are thus left with the following seven rows that are each consistent with all three given conditions (you'll notice that *seven* rows is, itself, consistent with the condition that the hunters occupied their camp for *seven* days, with one row for each day since we were told that no two days were identical):

A	B	C	D
0	0	0	0
0	0	1	0
0	1	0	1
1	0	0	0
1	0	0	1
1	0	1	0
1	0	1	1

So, the answer to Puzzle 4 is immediately established: D hunted on three days, once with B alone, once with A alone, and once with both A and C. These seven rows also provide the answers to the same questions for A, B, and C (for example, A hunted on four days, B hunted on just one day, and C hunted on three days, and, you'll notice, there was one day—the row of all 0s—during which *nobody* hunted).

To finish my discussion of logic puzzles, let me show you one that Shannon himself cited in one of his papers, a problem *so* hard that I think it *will* turn your brain to mush. It did mine!

> It is known that salesmen always tell the truth and engineers always tell lies. B and E are salesmen. C states that D is an engineer. A declares that B affirms that C asserts that D says that E insists that F denies that G is a salesman. If A is an engineer, how many engineers are there?[4]

There are seven Boolean variables here (A through G), and so there are potentially a total of $2^7 = 128$ salesman/engineer combinations. Many are immediately eliminated, of course, because of the given conditions. For example, since A is stated to be an engineer, half of the 128 initial possibilities are eliminated and we are down to 64. We eliminate even more using the facts that B and E are both given as salesmen. Each condition eliminates half again, and so we end-up with just 16 combinations to consider. And finally, we can eliminate half again using "C states D is an engineer." After all, C is either an engineer or a salesman. If an engineer, he lies, and so D is actually a salesman. If a salesman, he tells the truth, and so D is an engineer. In either case, $C \neq D$, and imposing that requirement leaves us with just 8 combinations.

But, if you can wade through "A declares that B affirms that C asserts ..." then you have done better than I have been able to do. Shannon doesn't provide a solution, either, but the paper he got it from does[5]. In fact, there are four combinations that are consistent with all the given conditions. Using the notation A and $\bar{A}$ to denote "A is an engineer" and "A is a salesman," respectively, the given solutions are: $A\bar{B}\bar{C}D\bar{E}\bar{F}G$, $A\bar{B}\bar{C}D\bar{E}F\bar{G}$, $A\bar{B}C\bar{D}\bar{E}\bar{F}G$, and $A\bar{B}C\bar{D}\bar{E}F\bar{G}$. In all four solutions, there are exactly three uncomplemented variables, and so that is the answer: there are three engineers. The authors, like Shannon, also do not provide a paper derivation, but rather give the circuit diagram of a relay computer that produced the four results. They do provide one hint: they observe that "A states $\{X\}$" is equivalent to "A is an engineer *or else* $\{X\}$ is true," and they use that equivalence to explain their computer circuit. How to use it in a paper derivation, however, has escaped me. Perhaps a reader will have more success —if so, please send me your derivation!

4.6 VISUALIZING BOOLEAN FUNCTIONS

At this point you have all the pure technical background you need to understand the *mathematics* of combinatorial digital circuits. There is, however, one last topic that, while it adds little or nothing to the theory, tremendously adds to the *engineering* of combinatorial digital circuit design. This innovation reached its modern form in present-day textbooks at the relatively recent date of 1953, in an influential paper published by the American physicist Maurice Karnaugh (born 1924)[6]. Karnaugh's so-called *map method* allows the pictorial representation of a Boolean function, which, in turn, allows the easy construction of alternative mathematical forms for the function—alternatives that may, in some way, be "better" in an engineering sense than other forms. Karnaugh's maps are intimately connected to the eighteenth-century Euler diagrams (see Figure 4.2.1 again) and their nineteenth-century descendent called *Venn diagrams*, named after the English philosopher John Venn (1834–1923).

To start, consider the following truth table for the Boolean function $F(A, B)$:

A	B	F
0	0	1
0	1	0
1	0	1
1	1	1

Now, from any truth table we can write the combinatorial Boolean function it represents by, for each 1 entry in the function column, including a logical product term with each input variable represented *if we complement* that is, *negate* an input variable if it has the value 0. Doing that, we get the Boolean function in the form of a "sum of products"—the so-called SOP form. In particular,

$$F = \bar{A}\bar{B} + A\bar{B} + AB. \tag{4.6.1}$$

We can algebraically simplify this in a couple of ways:

$$F = (\bar{A} + A)\bar{B} + AB = \bar{B} + AB \tag{4.6.2}$$

or

$$F = \bar{A}\bar{B} + A(\bar{B} + B) = \bar{A}\bar{B} + A. \tag{4.6.3}$$

All three of our expressions for F are logically equivalent. The Karnaugh map for F is shown in Figure 4.6.1, which plots a 1 in each sub region that is "covered" by the terms in (4.6.1). The logical equivalence of (4.6.2) and (4.6.3) is "visually proven" by observing that the maps of each are the same as Figure 4.6.1.

But notice, carefully, that Figure 4.6.1 also shows that there is a another way to write F that is not derivable algebraically with the ease

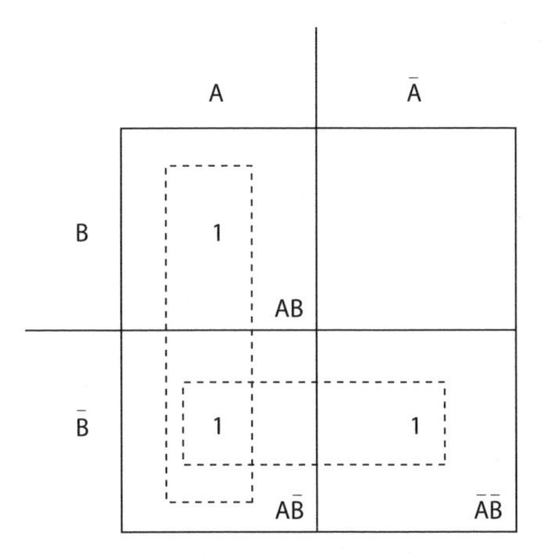

Figure 4.6.1. The Karnaugh map for F

that we got (4.6.2) and (4.6.3):

$$F = A + \bar{B} \tag{4.6.4}$$

The two terms of (4.6.4) "cover" one of the subsquares twice (the $A\bar{B}$ square), yes, but in Boolean algebra $1 + 1 = 1$, and so multiple coverings of a subregion can occur with no problem caused. All *four* of our expressions for F look quite different from one another but yet are logically equivalent.

Next, consider the following truth table for the Boolean function $G(A, B, C)$:

A	B	C	G
0	0	0	1
0	0	1	1
0	1	0	0
0	1	1	0
1	0	0	0
1	0	1	1
1	1	0	0
1	1	1	0

From the table we have

$$G = \bar{A}\bar{B}\bar{C} + \bar{A}\bar{B}C + A\bar{B}C, \tag{4.6.5}$$

which gives the three-variable Karnaugh map shown in Figure 4.6.2.

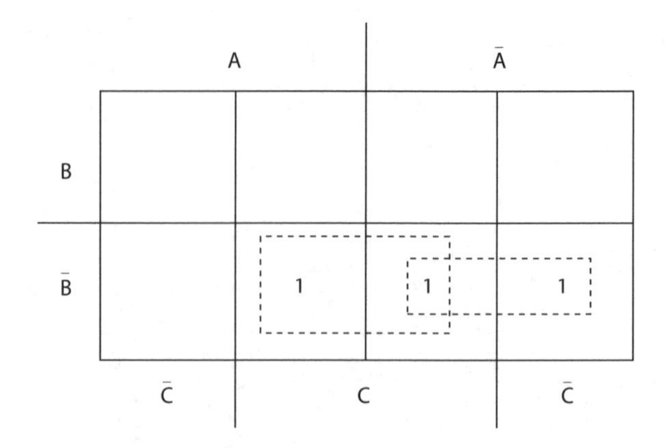

Figure 4.6.2. The Karnaugh map for G

The map immediately shows the following alternative form for G:

$$G = \bar{A}\bar{B} + \bar{B}C. \tag{4.6.6}$$

Perhaps even more interesting is that the map immediately allows us to write an expression for $\bar{G}$, by simply looking at the inverse (complementary) map obtained by writing a 1 everywhere there isn't a 1 in the G-map, as shown in Figure 4.6.3. Thus,

$$\bar{G} = A\bar{C} + B \tag{4.6.7}$$

and, using De Morgan's theorem,

$$G = \overline{(A\bar{C} + B)} = \overline{A\bar{C}}\,\bar{B},$$

or, again using De Morgan's theorem,

$$G = (\bar{A} + C)\bar{B}. \tag{4.6.8}$$

Notice that (4.6.8) is in the form of a logical "product of sums" — the so-called POS form—as opposed to the SOP form in (4.6.5) and (4.6.6). Depending on circumstances, one form may be preferred over the other for engineering reasons.

To finish this section, let me show you one last use of a Karnaugh map that easily accomplishes what can be a very difficult task if

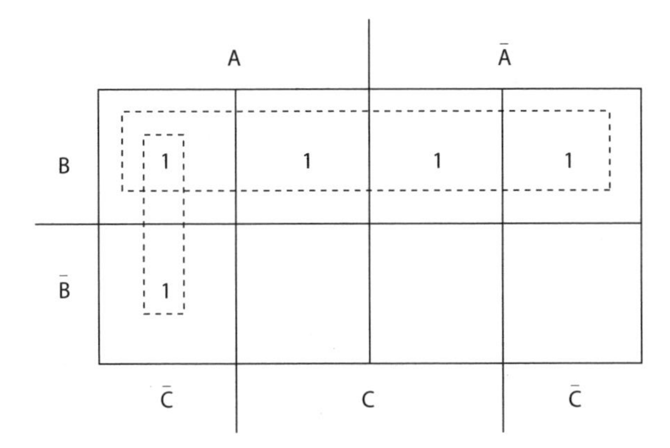

Figure 4.6.3. The Karnaugh map for $\bar{G}$

attempted analytically. Suppose one has the following truthtable for the four-variable Boolean function $H(A, B, C, D)$:

A	B	C	D	H
0	0	0	0	0
0	0	0	1	0
0	0	1	0	0,1
0	0	1	1	0
0	1	0	0	0
0	1	0	1	1
0	1	1	0	0
0	1	1	1	0,1
1	0	0	0	0,1
1	0	0	1	1
1	0	1	0	0,1
1	0	1	1	0
1	1	0	0	1
1	1	0	1	1
1	1	1	0	0
1	1	1	1	1

This truth table probably looks a bit odd to you; you may be asking, "Why are there rows in the table where the entry for H is not just a 0 or just a 1, but rather *both*?" They are there because when the input

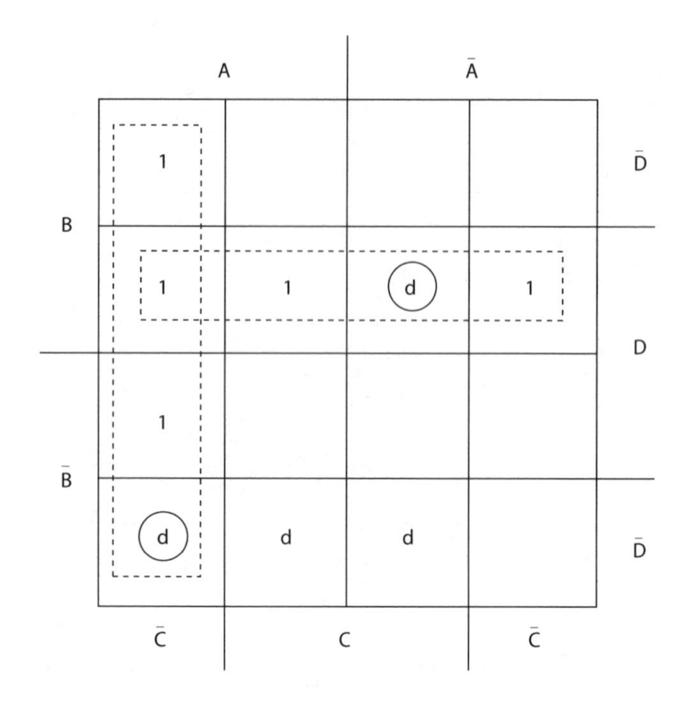

Figure 4.6.4. The Karnaugh map for H

variables have the values shown in each of those particular rows, we *don't care* what the output function value is! Now, of course, once we learn how to actually construct electronic circuits that realize Boolean functions in hardware, then there will be a specific output value for every possible combination of inputs. So, the output function H must have just one or the other value (0 or 1, but not both!) for each and every row, but we are free to pick, independently, either output value for each of the "don't care" rows. Our choice will be guided by what makes the final expression for H the "simplest" by some criterion.

To do our analysis, I'll first write the terms that give the value of 1 for H, and then (in parentheses) I'll write the "don't care" terms. Thus,

$$\begin{aligned} H &= \bar{A}B\bar{C}D + A\bar{B}\bar{C}D + AB\bar{C}\bar{D} + AB\bar{C}D + ABCD \\ &\quad + (\bar{A}\bar{B}C\bar{D} + \bar{A}BCD + A\bar{B}\bar{C}\bar{D} + A\bar{B}C\bar{D}). \end{aligned} \tag{4.6.9}$$

Figure 4.6.4 is the Karnaugh map for H, with the "don't care" terms plotted as d's (Karnaugh's original notation). With Figure 4.6.4

displaying the interaction of the "don't care" terms with the required terms, I think the simple form of

$$H = A\bar{C} + BD \tag{4.6.10}$$

figuratively jumps off the page, where I've taken just two of the *d*'s (the two circled inside the dashed lines) as 1s.

What if we have Boolean functions of more than four input variables? Karnaugh had some suggestions for such cases, one involving three-dimensional maps, but as far as I know none have caught on with professional logic design engineers. Four variables are the most I've ever seen people use with ease; beyond that, other techniques (such as computer-aided algorithmic codes) are the simplification tools of choice.[7] At the end of his paper Karnaugh wrote, "Beyond nine variables, the mental gymnastics required ... will, in general, be formidable." I think that's true, but also that experience has shown that Karnaugh's upper limit of nine variables was far too optimistic.

NOTES AND REFERENCES

1. Theodore Hailperin, "Boole's Algebra Isn't Boolean Algebra," *Mathematics Magazine*, September 1981, pp. 173–184.

2. T. A. Broadbent, "George Boole (1815–1864)," *Mathematical Gazette*, December 1964, pp. 373–378.

3. De Morgan—who you'll recall from Chapter 3 was the immediate cause of Boole extending his mathematical writings into logic— published his work in 1859, in the *Cambridge Philosophical Transactions*. Boole and De Morgan carried on a quite interesting correspondence for years that lasted until Boole's death. Many of their letters have survived (but even more, alas, have not), and reading them gives much insight into the character of each man. You can find their letters in G. C. Smith, *The Boole–De Morgan Correspondence 1842–1864*, Oxford University Press, 1982. A quite interesting analysis of the role of spirituality in the mathematics of both Boole and De Morgan is in the book by Daniel J. Cohen, *Equations from God: Pure Mathematics and Victorian Faith*, Johns Hopkins University Press 2007, pp. 77–136.

4. C. E. Shannon, "Computers and Automata," *Proceedings of the IRE*, October 1953, pp. 1234–1241.

5. D. M. McCallum and J. B. Smith, "Mechanized Reasoning: Logical Computers and Their Design," *Electronic Engineering*, April 1951. The puzzle in this paper is as I've stated it here, not as in Shannon's paper, which has a misleading typo.

6. M. E. Karnaugh, "The Map Method for Synthesis of Combinatorial Logic Circuits," *Transactions of the American Institute of Electrical Engineers* 72 (Part 1, Communication and Electronics), November 1953, pp. 593–599.

7. When I was a digital system designer in the 1960s, I didn't worry very much about minimization. If a simplification was obvious, then of course I used it. But I didn't try to do anything "clever." In college, minimization was a big deal, and textbook writers generally made minimization a central topic. In real life, however, the fact is that right up until the project is actually accepted, shipped, and paid for by the client, designers are ever alert for changes, that is, memos (called *engineering change orders*) from the client saying something like "we didn't really mean what we originally asked for in paragraph 19 on page 73, but instead we now want . . ." To make such changes on a machine that is already 80% wired on the production floor meant that you'd better have some spare logic gates in your design (typically, 10% of the nominal design). Minimization was actually counterproductive, something I didn't learn until I actually designed for a paycheck and not for homework points.

5

Logic Switching Circuits

In his master's thesis he showed how an algebra invented in the mid-1800s by the British mathematician George Boole—which deals with such concepts as "if X or Y happens, and not Z, then Q results"—could represent the workings of switches and relays in electronic circuits. The implications of the paper were profound: engineers now routinely design computer hardware and software, telephone networks and other systems with the aid of Boolean algebra. Shannon downplays the discovery. "It just happened no one else was familiar with both those fields at the same time," he says. He adds, after a moment of reflection, "I've always loved that word, Boolean."

—In a profile of Claude Shannon, *Scientific American* (January 1990)

5.1 DIGITAL TECHNOLOGY: RELAYS VERSUS ELECTRONICS

Today's digital circuitry is built with electronic technology that the telephone engineers of the 1930s and the pioneer computer designers of the 1940s would have thought to be magic. And I mean that literally: to quote science fiction writer Arthur C. Clarke's famous third law: "Any sufficiently advanced technology is indistinguishable from magic."[1] An example of this is the ordinary radio, which while commonplace to us (modern kids probably find AM radio just a bit boring!) would have been magic to the greatest of the Victorian scientists, including James Clerk Maxwell, himself who first wrote the equations that give life to radio. In the Middle Ages such a gadget would have gotten its owner burned at the stake —what else, after all, could a "talking box" be but the work of the devil?

The first real digital technology took the form of electromagnetic relays in telephone switching exchanges. Then came vacuum tube digital circuitry (although the tube itself had been in radio circuits since before 1910), and then discrete transistors, and then integrated transistor circuits, and then .... The acronyms abound: DTL, TTL, ECL, CMOS, I^2L, and who knows what next year. The one thing that remains the same is the math, the Boolean algebra that is the central star of this book. As I stated in the introduction, we will *not* discuss electronics—I am an electrical engineer and so I hope you appreciate how much I am giving up here!—because any specific technology is simply secondary (or even tertiary) to the spirit of this book.

But to really appreciate what Shannon did with Boole's math, we of course have to discuss electrical circuits in *some* form, and so I've decided to go back to the beginning and use relays. These are intuitively easy-to-understand devices, and historically they are the technology that Shannon himself used in his switching analyses. To my knowledge, Shannon always wrote, until the end of his career, in terms of relays when discussing digital switching circuits. He of course knew of vacuum-tube and solid-state electronics—he was at Bell Labs when the transistor was invented there and was a friend of John Pierce, the Bell Labs electrical engineer who gave the new device its name (Shannon's second wife was a mathematician in Pierce's group at Bell Labs when they met)—but the *relay switch* was always the device of choice for Shannon. The brains of his 1950 maze-running mouse (see the photograph in Figure 3.2.1 again) were built from something like 75 relays. A note on the back of the original photograph specifically says Shannon built the mouse to "illustrate the capabilities of telephone relays."

5.2 SWITCHES AND THE LOGICAL CONNECTIVES

A *switch* is a mechanical device, either hand-operated (a toggle or a push button) or electrically actuated (a relay, which I'll discuss later

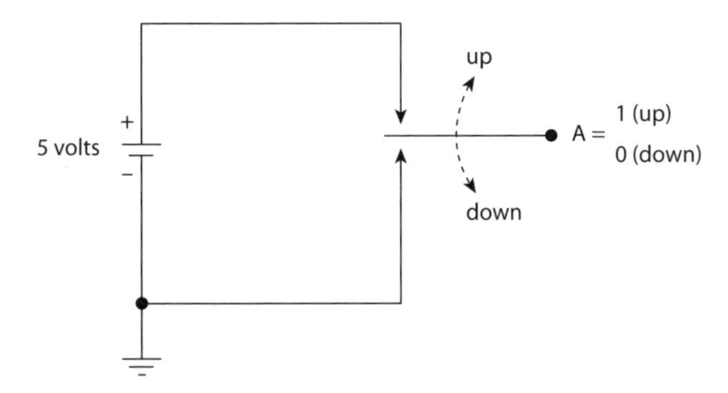

Figure 5.2.1. A hand-operated switch.

in this chapter, or electronic circuits, which I *won't* discuss) that, in its simplest form, we'll represent as a movable contact that can be flipped back and forth between two fixed contacts, as shown in Figure 5.2.1. In the figure I've labeled the movable contact A, and A is a Boolean variable. As the figure shows, the movable contact voltage is either +5 volts or 0 volts (ground), which I'll take here to represent logical 1 or logical 0, respectively. When the movable contact A is touching the upper fixed contact, we have $A = 1$, and when the movable contact A is touching the lower fixed contact, we have $A = 0$.

It was Shannon's insight to see that putting switches in series or parallel allows one to construct logical *and* or logical *inclusive-or* electrical circuits, respectively.[2] And clever use of switch contacts allows the construction of the *not* logical operation, too. For example, in Figure 5.2.2 the lamp illuminates if switch A *and* switch B are logical 1, while in Figure 5.2.3 the lamp illuminates if switch A *or* switch B (*or both*) are logical 1. You'll notice that in Figure 5.2.3 the lower fixed contacts of A and B (the $\bar{A}$ and $\bar{B}$ contacts) are not connected to ground. If they were, then it would be possible to short the +5 volt power supply to ground, resulting in much pyrotechnics! [3]

In many applications it is convenient to have both A and $\bar{A}$ available at the same time, and the arrangement of Figure 5.2.4 provides that. There the +5 volt power supply is connected to the movable contact, and the two fixed contacts provide A and $\bar{A}$. When the movable contact touches either fixed contact, then that fixed contact voltage is, of course, +5 volts (logical 1). The other fixed contact is connected to

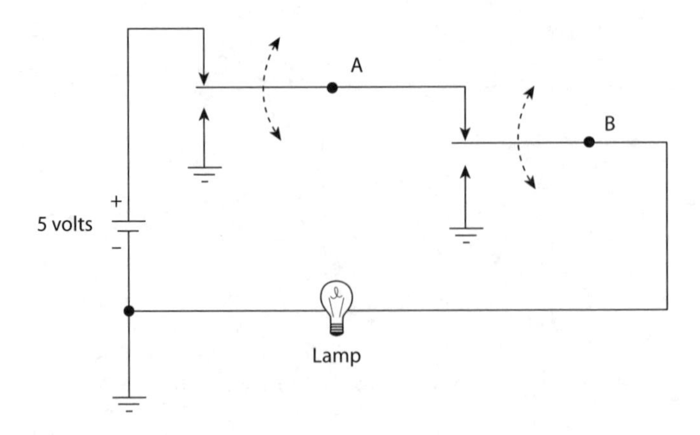

Figure 5.2.2. *Series* means *and*.

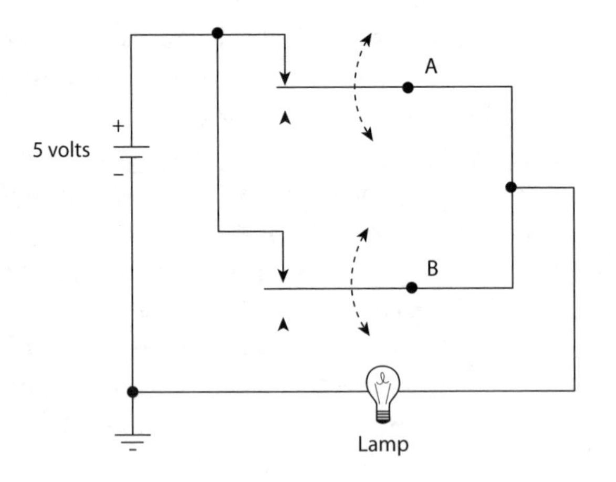

Figure 5.2.3. *Parallel* means *or*.

ground (0 volts, or logical 0) through a resistor R. Without those resistors the isolated contact would be electrically "floating," and its voltage with respect to ground would be undefined. The resistors provide electrical paths to ground; they are called *pull-down resistors* (they *pull* the voltage of an otherwise floating contact *down* to ground potential). They also prevent the power supply from being shorted to ground when the movable contact is at either fixed contact. The circuit of Figure 5.2.5 shows how to wire two lamps so that lamp 1 illuminates if $A = 1$ *and* $B = 1$, while lamp 2 illuminates if $A = 1$ *and* $B = 0$ ($\bar{B} = 1$), *or* if $A = 0$ ($\bar{A} = 1$).

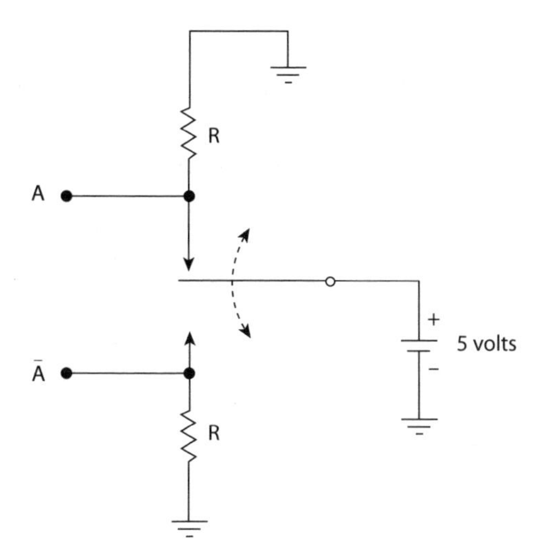

Figure 5.2.4. One way to generate **A** and **Ā**.

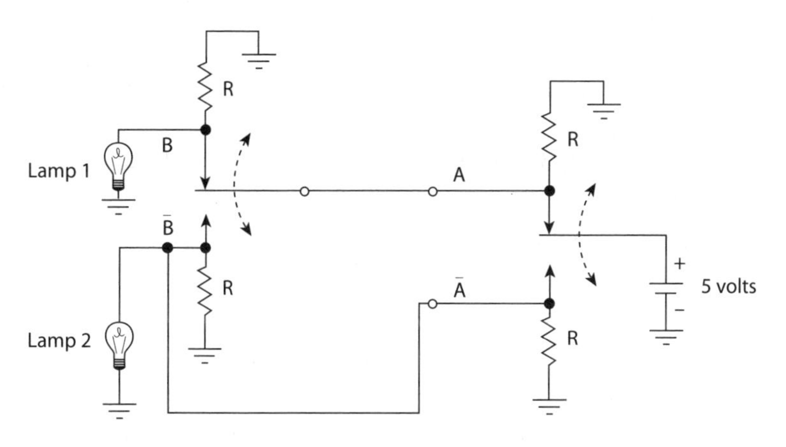

Figure 5.2.5. A two-lamp circuit.

5.3 A CLASSIC SWITCHING DESIGN PROBLEM

Imagine that there is a wall sconce lamp halfway down a staircase, which can be controlled from either of two switches, one at the top of the stairs and one at the bottom. How are the lamp and the switches connected with the power supply? That is, one can turn the light **on** or **off** from either switch, independent of the setting of the other switch. This is a situation that occurs in probably most homes in the modern world and yet, when first encountered, most people have trouble "seeing" how to do it. If we define the two switches by the Boolean variables A and B, and if we define the lamp by the Boolean variable L, with $L = 1$ meaning the lamp is **on**, and $L = 0$ meaning the lamp is **off**, then a truth table for the circuit logic is

A	B	L
0	0	0
0	1	1
1	0	1
1	1	0

The explanation for this table is straightforward. We define the lamp to be **off** when both switches are in their 0 positions. Then, changing *one* of either of the two switches should turn the lamp **on** (that, in going from the 0 0 row in the table to either of the rows 0 1 or 1 0). Then, changing *one* of either of the two switches once more should turn the lamp **off** (that is, in going from the 0 1 row to either of the rows 0 0 or 1 1, or going from the 1 0 row to either of the 0 0 or 1 1 rows). From the table, then, we have

$$L = \bar{A}B + A\bar{B}. \tag{5.3.1}$$

This is not the only solution, however, as you should be able to see that the following truth table also satisfies the requirements of the problem (the result of defining the lamp to be **on** when both switches are in their 0 positions, rather than **off**):

A	B	L
0	0	1
0	1	0
1	0	0
1	1	1

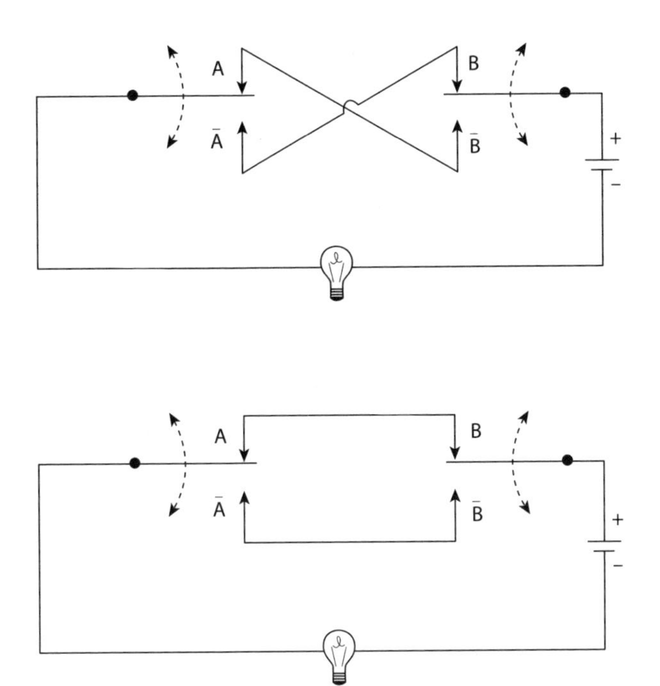

Figure 5.3.1. Two solutions to the staircase light problem.

and so now

$$L = \bar{A}\bar{B} + AB. \tag{5.3.2}$$

Figure 5.3.1 shows the resulting circuits (the top circuit for (5.3.1) and the bottom circuit for (5.3.2)). You should be able to see that either circuit solves our staircase problem. There is no reason to prefer one circuit over the other and so, for a problem that often seems quite puzzling on first encounter, we see that there are in fact two solutions, and Boolean algebra easily shows us both of them.

5.4 THE ELECTROMAGNETIC RELAY AND THE LOGICAL NOT

So far, our development of electrical circuits that implement the various logical switching functions may seem a bit ad hoc to you.

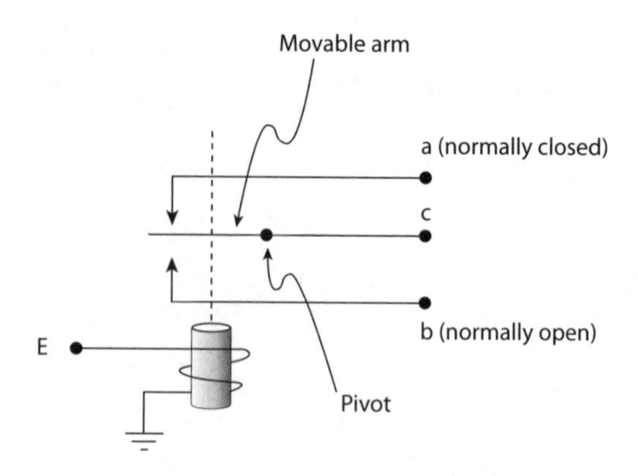

Figure 5.4.1. An electromagnetic relay.

Sometimes the fixed contacts are grounded, other times not, sometimes the power supply voltage is applied to the fixed contacts, other times to the movable contact. Indeed, before Shannon, the design of switching circuits was more of an art form than it was engineering science. Each new problem demanded a new, distinct invention. What is needed is a set of standard building block circuits—*logic gates* —that one can routinely interconnect in a straightforward way to construct any combinatorial logical switching circuit desired. What we need is the logic equivalent of an erector set using many copies of a small number of fundamental components. In this section, and in the next, I'll show you how that can be done with relays.

Today, of course, logic gates are electronic in nature, as I discussed in the opening section of this chapter, but the relay hung on for quite a while. With the invention of the transistor in 1948, and the consequent tremendous reduction in size, speed, and power requirements of logic gates, the days of the relay were clearly numbered. Nevertheless, once could still find, as late as the mid-1950s, college-level texts devoted to teaching switching circuit design using only large, slow, power-hungry relay technology.[4]

To begin our development of relay logic gates, consider Figure 5.4.1, which shows how simple a device is the electromagnetic relay switch. It consists of a coil of wire around a piece of iron, with one end of the coil grounded and the other end connected to a voltage E. When the current-carrying coil is sufficiently energized (by E being

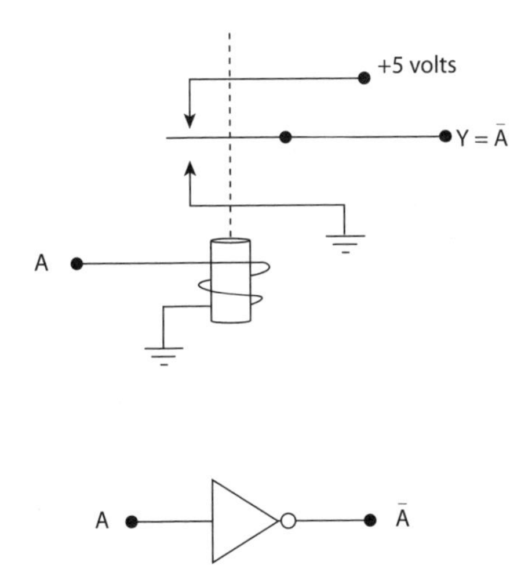

Figure 5.4.2. The relay logical inverter (NOT gate).

sufficiently large), then the iron is an electromagnet with a magnetic field strong enough to attract a spring-tensioned arm away from its "normal" position to a new position. When the energizing current is removed, the spring tension pulls the arm back to its normal position.

When the coil is carrying a current that exceeds some minimum value the relay's movable arm switches from the upper contact (a) to the lower contact (b). Contact a is said to be the n.c. or *normally closed* (coil *not* energized), or *break* contact. Contact b is said to be the n.o. or *normally open*, or *make* contact. When the coil is not energized, we have an electrical path through the break contact a and terminal c, and when the coil is energized we have an electrical path through the make contact b and terminal c.

To be specific, let's assume our relay has a coil resistance of 1,000 ohms, and that the coil is energized when the coil current exceeds 0.004 amperes (that is, 4 milliamperes = 4 ma). Finally, let's say a voltage of zero volts (ground) is logical 0, and that +5 volts is logical 1. So, when we apply logical 0 ($E = 0$) to the relay coil, the coil current is zero (<4 ma) and the relay is *not* energized, while when we apply logical 1 ($E = 5$ volts) to the relay coil, the coil current is 5 ma (>4 ma) and the coil *is* energized.

We can now easily build our first logic gate, the NOT (inverter) gate shown in Figure 5.4.2. You should be able to see that $Y = 5$ volts

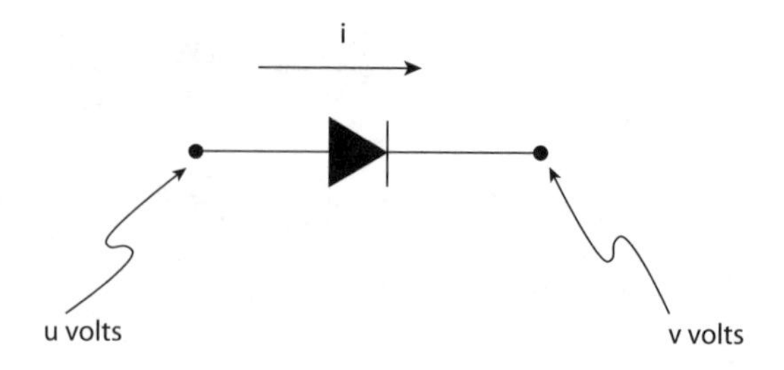

Figure 5.5.1. The diode.

when $A = 0$ volts, and that $Y = 0$ volts when $A = 5$ volts. That is, A and Y are each Boolean variables such that $Y = \bar{A}$. Rather than having to reproduce the intricate circuit details of the logical inverter over and over when drawing complicated combinatorial logic diagrams, the standard logic symbol for the inverter gate is shown directly below the relay circuit in the figure.

5.5 THE IDEAL DIODE AND THE RELAY LOGICAL AND and OR

To understand the operation of relay AND and OR gates, we need to take a quick detour and bring the *diode* into the discussion. Diodes have been a part of electrical engineering since the earliest days of radio,[5] and even earlier, but for many years the physics behind what is often called *diode action* was a mystery. To really get into what happens inside a modern solid-state diode requires some discussion of electronics and quantum mechanics, but for our purposes here some very simple imagery will do.

A diode is what electrical engineers call a *voltage-controlled switch*, and they use the symbol shown in Figure 5.5.1 when drawing circuits. When the diode voltages u and v, are such that $u > v$, then the diode presents a low resistance (in the ideal limit, zero resistance). The diode is said to be *forward biased*, and the current in the diode (flowing in the direction of the arrowhead) is determined by the circuitry that surrounds the diode. The voltage drop across a forward-biased diode is small (in the ideal limit, the drop is zero). When the diode voltages

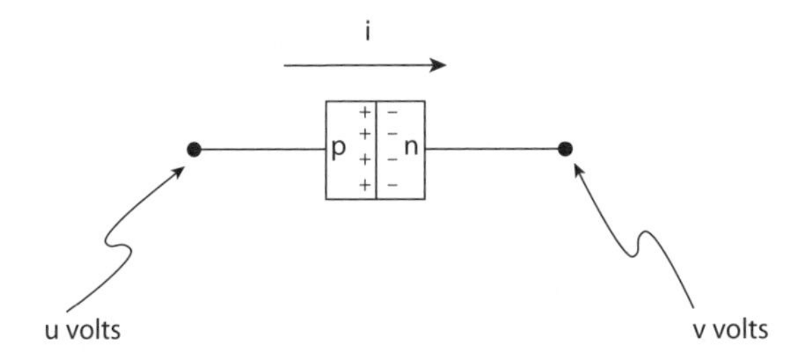

Figure 5.5.2. The pn-junction.

u and v are such that $u<v$, then the diode presents a high resistance (in the ideal limit, infinite resistance). The diode is said to be *reverse biased*, and the current in the diode is zero. (Think of the vertical line in the diode symbol as a brick wall that stops any current flow opposite the arrowhead!)

You can understand how diode action occurs as follows. Imagine that there are two kinds of material, called p-stuff and n-stuff. Both kinds of material can actually be made, in fact, by adding (in a delicate process called *doping*) certain impurities to pure silicon. Exactly what occurs when doping is done is where the quantum mechanics I mentioned earlier comes in, but you can simply think of p-stuff as acting like a conducting material in which the carriers of electrical charge are positive (what physicists and electrical engineers call *holes*) and n-stuff as acting like a conducting material in which the carriers of electrical charge are negative (the well-known *electrons*). Now, imagine a slab of p-stuff joined to a slab of n-stuff, as shown in Figure 5.5.2. The two slabs form what is called a *pn-junction*, and it is the basis for all of modern solid-state electronics. In his 1987 *Omni* interview, Shannon declared the transistor to be "the most important thing discovered this century." We'll be satisfied here, however, to see how the pn-junction explains diode action.

Imagine that we apply a positive voltage u and a negative voltage v to the pn-junction terminals. Then, because like polarities repel, the positive u pushes the positive holes across the junction from the p-stuff, into the n-stuff, and the negative v pushes the negative electrons across the junction from the n-stuff into the p-stuff. The result, from both pushes and resulting charge carrier motions, is a current $i > 0$ directed from p-stuff to n-stuff and we have a forward-biased diode. More generally, it isn't really necessary that $u > 0$ and $v < 0$, just that $u > v$, (for example, $u = -20$ volts and $v = -30$ volts would result in a forward-biased diode).[6]

If, on the other hand, $u < 0$ and $v > 0$, then, since unlike polarities attract, the negative u pulls the positive holes *away* from the junction, and the positive v pulls the negative electrons *away* from the junction. The result, from both pulls, is that no charge carriers flow across the junction, and so $i = 0$ and we have a reverse-biased diode. More generally, it isn't really necessary that $u<0$ and $v>0$, just that $u<v$ (again, see note 5).

You should now be able to understand how the relay circuits of Figure 5.5.3 implement the logical inclusive-OR and the logical AND operations. In the upper circuit (the inclusive-OR), if both A and B are at ground potential (both are logical 0), then the coil voltage E is obviously at ground potential as well, and so the relay is not energized. Thus, the output Y is connected through the break contact to ground, and so Y is logical 0. If, however, A is +5 volts (logical 1) while B is at ground potential (logical 0), then diode D1 is forward-biased and E becomes +5 volts which both energizes the relay and reverse biases diode D2. (The presence of D2 is necessary to prevent A from shorting directly to ground via B.) Thus, with the relay energized, Y is connected through the make contact to +5 volts, and so Y is logical 1. The same argument applies to the case of A as logical 0 while B is logical 1 (except now D2 is forward biased and D1 is reverse biased). And finally, if both A and B are logical 1, then both diodes are forward biased, and so Y, is logical 1. Thus, in summary,

$$Y = A + B.$$

The behavior of the lower circuit (the AND) is only slightly more involved. If A and B are both at ground potential (both are logical 0), then current flows from the +5 volt power supply through the resistor r (typically small, on the order of 100 ohms, and so negligible compared to the resistance of the relay coil) to *ground* via A and B because both D1 and D2 are forward biased. That is, E is at ground potential, too, and so the relay is not energized. Thus, Y is connected

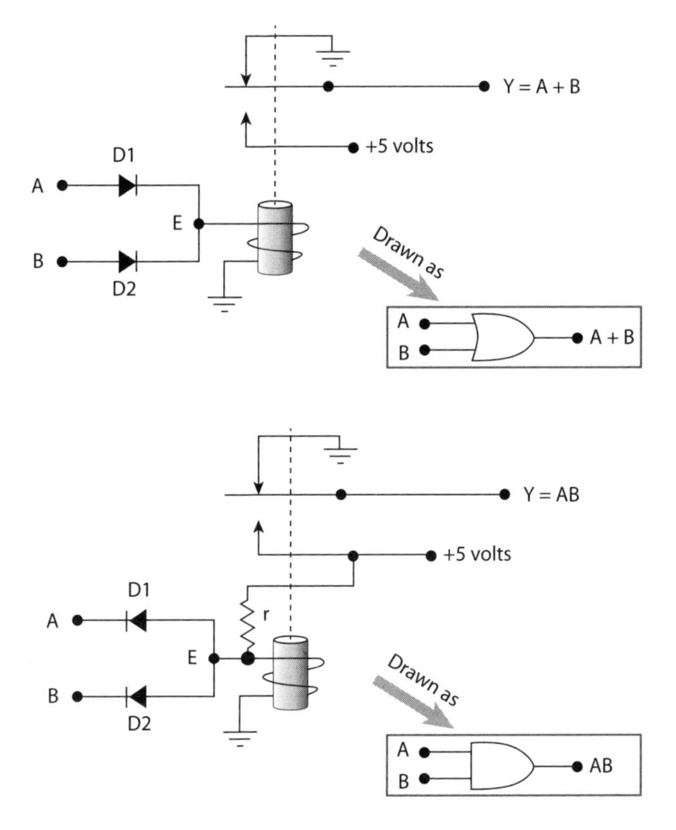

Figure 5.5.3. The relay inclusive-OR and AND logic gates.

to ground (logical 0) through the break contact. If A is at +5 volts while B is at ground, then D2 is forward biased and so the relay remains unenergized. Similarly for A at ground and B at +5 volts. When both A and B are at +5 volts (when both A and B are logical 1), then current still flows from the +5 volt power supply, but now through the relay coil; that's because both diodes have +5 volts on their left terminal and voltage E (which is less than +5 volts because of the voltage drop across r) on their right terminal, and so both diodes are reverse biased. The relay coil is the only path left for the current. Thus, the relay is energized and so Y is connected to +5 volts via the make contact. So, Y is logical 1 only when A is logical 1 *and* B is logical 1, and so

$$Y = AB.$$

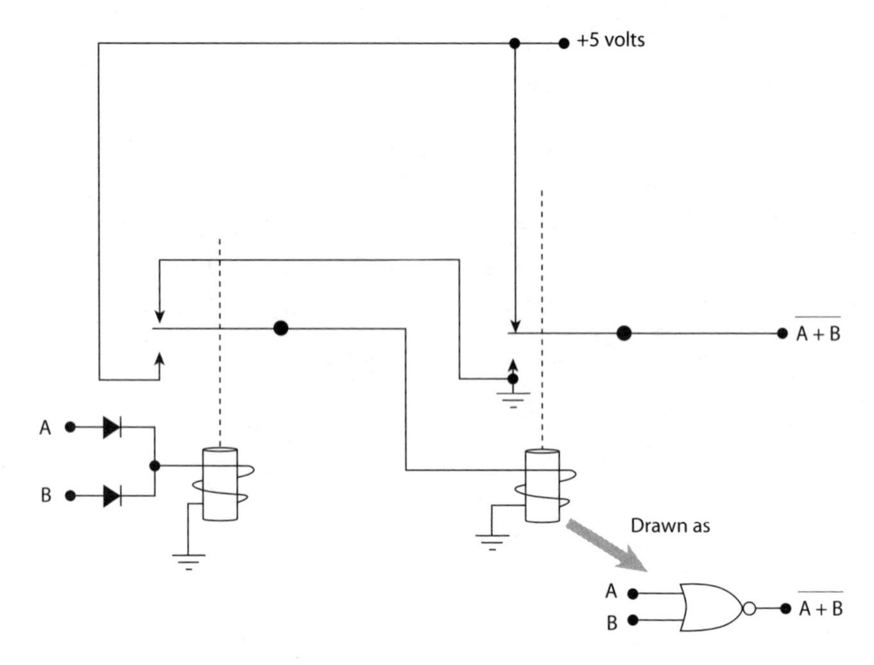

Figure 5.5.4. The relay NOR logic gate.

With these three kinds of logic gates one can construct any combinatorial circuit. In actual practice, the three different gate types are not generally present; in the interest of standardization and economy, one finds that all combinatorial logic is built either from just not-AND gates or from just not-Inclusive-OR gates (the NAND or the NOR, respectively). To build either the NAND or the NOR with relays is now, of course, trivial for us; the relay NOR is shown in Figure 5.5.4. In Figure 5.5.5 you can see how easy it is to go the other way; that is, to build the NOT, Inclusive-OR, and AND functions from just NOR gates (the inputs with logic 0 are connected directly to ground). As an exercise, you should try your hand at doing the same with just NAND gates.

With standardization we have not only a single logic gate function out of which everything is built, but also the reality that the standard logic gate (here, the NOR) also has a fixed number of inputs. In the circuits I've shown you so far, all logic gates have two inputs. But suppose we need (for whatever reason) a three-input NOR? How can

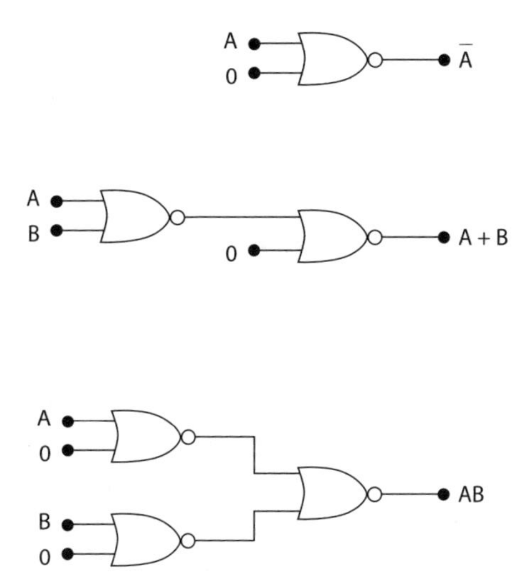

Figure 5.5.5. NOT, Inclusive-OR, and AND logic gates from just NOR gates.

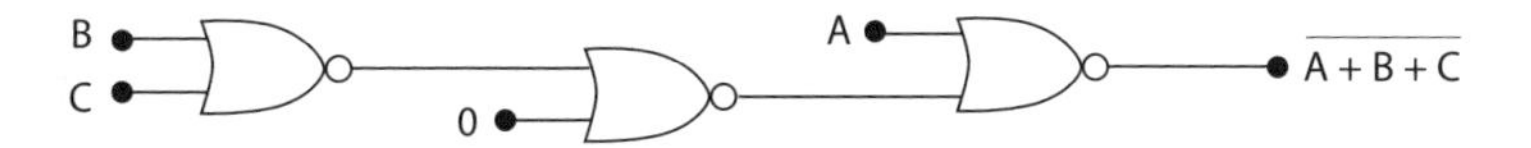

Figure 5.5.6. Building a 3-input NOR from 2-input NORs.

we do that? The circuit in Figure 5.5.6 shows how to make a 3-input NOR using just 2-input NOR gates.

5.6 THE BI-STABLE RELAY LATCH

I'll finish this chapter by giving you an example of how one can use relays to build a new type of logic circuit that is a revolutionary step beyond "mere" combinatorial circuitry. It is, in fact, central to the construction of all modern digital machines. This new circuit will respond to one or more inputs, just as do combinatorial circuits, but it will have one additional capability: *memory*. After the inputs are removed, this new circuit —called a *latch*—will remain in one of two stable states, hence the name *bi-stable*, with each possible state characteristic of the specific nature of the now (perhaps) long-gone

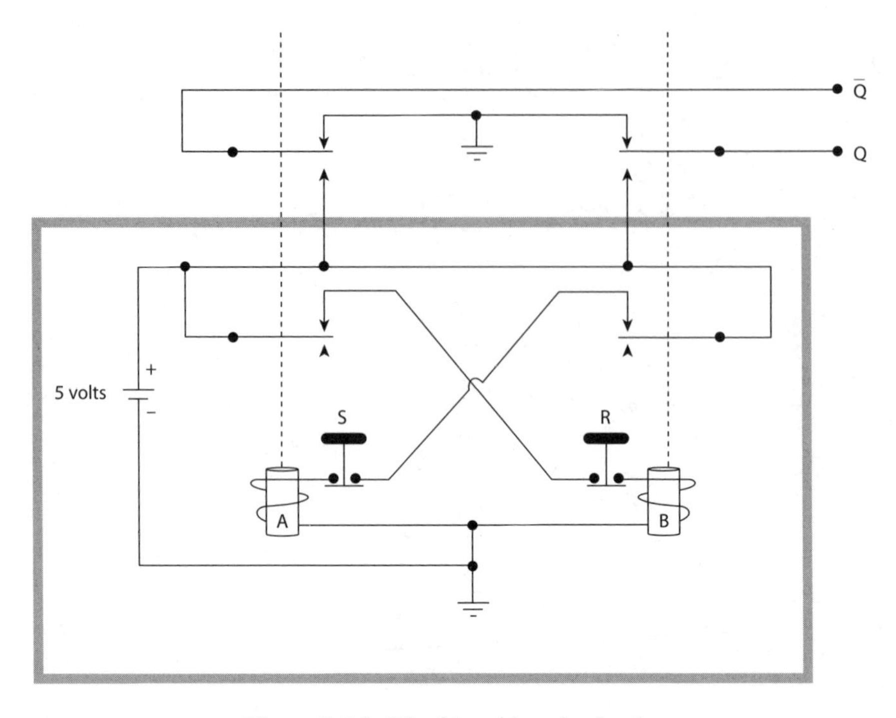

Figure 5.6.1. The bi-stable relay latch.

input(s). The state of the circuit can be changed with new input(s), but, until those new inputs arrive, the present state is remembered by the circuit. More picturesquely, the circuit can *flip* from one state to the other and then, later, *flop* back to the first state. The bi-stable latch is a rudimentary form of what is called a *flip-flop*; a true flip-flop is a latch with the additional input of a periodic clock signal that determines when state changes can occur. I'll say more about clocked latches in Chapter 8.

The bi-stable latch consists of two cross-coupled relays, as shown in Figure 5.6.1. These two relays (A and B) each have two sets of independent make-break contacts. The latch has two complementary outputs, denoted by Q and $\bar{Q}$. For each relay, one of its two sets of contacts is devoted to the actual operation of the circuit, and the other set is used to generate an output (relay A generates $\bar{Q}$ and relay B generates Q). Here's how the latch works.

To start, fix your attention on the circuitry inside the gray outline. Switches S and R are each normally closed push-buttons, as shown. When power is first applied, the coils of *both* relays are energized. However, no two relays are perfectly identical, not even if they come off the same production line, one right after the other. One relay will be faster than the other and respond to its energized coil first. For the sake of argument, let's say the faster of the two relays is B, although as you'll soon see our final conclusions will be independent of this arbitrary assumption. So, the break contacts on B open before the break contacts on A do, which breaks the coil current path to A, and so the break contacts for A remains in place. That keeps B's coil energized, and this is a stable state until something new occurs.

That "something new" happens when we push (that is, open) switch R (pushing switch S would have no impact because there already is no current in A's coil). That removes the coil current to B and that causes the spring-tensioned arms to return to B's break contacts, which energizes A's coil. That in turn causes A's break contacts to open and *that* further open-circuits the current path to B's coil. (Note that even with relays, a slow technology compared to electronics, this chain of events occurs in a fraction of a second, certainly less than 100 milliseconds.) And that means, even when we remove our finger and let push-button R return to its normally closed position, relay B remains without coil current, and A's coil current continues to flow through B's break contact, and so, again, we have a stable state until "something new" occurs. That "something new" happens when we push switch S (pushing switch R would have no impact) and the latch returns to the other stable state.

You'll notice that when A is energized (and so B is not) that the other sets of contacts for A and B are such that $\bar{Q} = 1$ and $Q = 0$, while when A is not energized (and so B is) we have $\bar{Q} = 0$ and $Q = 1$.

In summary, pushing R turns A **on** and B **off**, while pushing S turns A **off** and B **on**. We call A **on**, B **off** the *Reset* state ($Q = 0$, $\bar{Q} = 1$), and A **off**, B **on** the *Set* state ($Q = 1$, $\bar{Q} = 0$). Switches S and R do not have to be push-buttons, but instead might be other relays whose coils are energized by the output voltages from other logical circuitry, that is, other logic gates and/or other latches. As a final comment on the latch for now, it realizes the smallest possible memory storage unit of a *single bit*.

Now, to end this chapter on a cautionary note, there *is* a problem (in addition to the ones of speed, size, and power that I mentioned earlier) with using relays with their physically moving mechanical parts, instead of using electronics. You'll notice that in both Figure 5.5.4 (the relay NOR) and Figure 5.6.1 (the relay latch) we have the coil current of one relay flowing through the break contact of another relay. The many turns of wire that *are* relay coils have the property of *inductance*, which is the electrical analog of the mechanical property of all masses that we call *inertia*. What that means is that if a current exists in a relay coil, then that current "resists" any attempt at change; if you try to suddenly stop a coil current by opening the break contact of the relay that is carrying the current, a *spark*[7] will appear across the space between the break contact and the moveable contact. The spark is the current's attempt to continue to flow. Sparks are very intense, and very hot, hot enough to cause tiny spots of melting (*pitting*) on the contacts. (If you operate relay circuits in a dark room, you can often see the tiny flashes of light from the contact sparking.) After being subjected to repeated switching, relay contact sparking can cause sufficient accumulated damage to destroy the relay.[8] The result is what Shannon called a "crummy" relay, and in the next chapter I'll show you that this problem was of enough concern that Shannon devoted considerable effort to studying ways to counter it.

NOTES AND REFERENCES

1. For the curious, Clarke's first two laws are: "When a distinguished but elderly scientist states that something is possible, he is almost certainly right. When he states something is impossible, he is probably wrong," and (2) "The only way of discovering the limits of the possible is to venture a little way past them into the impossible."

2. Shannon's remark in the opening quotation, that he was the only one familiar in the late 1930s with both relay circuits and logical algebra, needs some qualification. In fact, the Japanese electrical engineer Akira Nakashima (1908–1970), while working for the Nippon Electric Company, started publishing in 1936 a series of papers quite similar to Shannon's work. The Russian physicist Victor Shestakov (1907–1987), who spent his entire career at Moscow State University, also made the relay switching interpretation

of Boolean algebra even earlier, in 1935. His work didn't appear until 1941, however, and like Nakashima's it passed unnoticed while Shannon's had immediate impact.

3. A 5-volt power supply might seem harmless—how could 5 volts electrocute anyone?—but that's not necessarily so. In even just a moderately large electrical machine the power supply may well be capable of providing a very large current (2,000 watts at 5 volts means 400 amperes). For that reason it was forbidden upon pain of being fired (when I was a logic designer many years ago) to work on the wiring of a machine while wearing hand jewelry. Accidentally shorting the power supply to ground through a ring or a watch would result in a current that could almost instantaneously turn the metal red hot, and you could easily lose a body part in the blink of an eye.

4. One such book was written in 1951 by three Bell Labs engineers, William Keister, Alistair E. Ritchie, and Seth H. Washburn: *The Design of Switching Circuits* (D. van Nostrand). It is my opinion that 99 out of 100 electrical engineers today (me included) would find 90% of its 556 pages to be as legible as if they were written in Martian. And yet, when I was in the electrical engineering program at Stanford in the late 1950s, I recall being told, while working one summer at a Southern California aerospace company, "Master this, son, and you'll be set for life." That from a senior electrical engineer while he held a copy of Keister, Ritchie, and Washburn in his hand. That was no doubt once true but, even as I heard those words more than fifty years ago, relay logic technology was going down the path into obsolescence that the typewriter, the slide rule, 8-track tape, 8 mm film, and the floppy disk would later follow. In fact, by 1954 Bell Labs was already finding it difficult to recruit young electrical engineers who could design relay logic circuits. Shannon's reaction that year was to coauthor an internal Bell Labs memorandum in which was described a relay circuit teaching tool for college laboratory instruction, with the goal of developing a pool of relay circuit designers ("It is suggested that if such a [relay] kit were developed and made available to colleges, it would materially aid our long-range policy toward cultivating switching engineers"). I don't think anything came from that suggestion. Relay logic books continued to appear for some years; for example, the 1954 French text by Rene A. Higonnet and Rene A. Grea, *Logical Design of Electrical Circuits* (translated into English and published by McGraw-Hill in 1958). When I told these stories to the students in a digital logic course I was teaching in 1972 at Harvey Mudd College, one of the students took my tales as a personal challenge. A couple of weeks later he appeared in class with a 4-bit relay counter (the input was provided by repeated operation of a push-button mounted on the counter chassis), built from discarded relays he found in various spare-parts bins in the Engineering Department basement. There was no electronics—the entire counter was simply wire and relays. I was suitably impressed, and I think Shannon would have been, too.

5. For the early history of diodes in radio, see my book *The Science of Radio*, Springer, 2001, pp. 53–65 and 102–106.

6. To understand this point, I need to introduce one additional concept—one not absolutely necessary to read this book but, since it's not really a difficult concept to grasp, why not?—that of the *electric field*. If two points separated by distance d have a voltage difference of Δv volts, then we say there is an electric field of $E = \frac{\Delta v}{d}$ volts/unit distance. The electric field has physical significance as the explanation for the force F that appears on an electric charge q when placed in the field: $F = qE = q\frac{\Delta v}{d}$, a force in the direction of the field E if $q > 0$ and opposite to field if $q < 0$. The electric field, like force, is a *vector field*; it has a magnitude and a direction (from the larger voltage to the smaller voltage). It is not the actual values of the two voltages, but their difference, that determines the electric field. Thus, positive holes feel a force in the direction of the field, and negative electrons feel a force opposite to the direction of the field. Both kinds of charges move in the field, but in opposite directions. Positive charges moving in one direction and negative charges moving in the opposite direction both represent positive current in the direction of the field. If the electric field in a pn-junction diode is from p-stuff to n-stuff, the charges move across the junction and we have current flow in a forward-biased diode; while if the electric field is from n-stuff to p-stuff, the charges move away from the junction and we have no current in a reverse-biased diode. Actually, there is a very small current because each of the p and n stuffs contain *very* low densities of what are called *minority charge carriers* (holes in p-stuff and electrons in n-stuff are *majority* charge carriers). That is, there are some electrons in the p-stuff and some holes in the n-stuff and they *do* move across the junction in a reverse-biased diode.

7. A spark is formed across an air gap when the electric field strength (see the previous note) exceeds something like 75,000 volts/inch. You might wonder how such a strong field can be created between the contacts in a relay circuit in which the power supply voltage is a mere 5 volts; a more advanced discussion than I've given here, of the mathematical physics behind what happens in an inductive circuit that is suddenly switched, is in my book *The Science of Radio* (see note 5), pp. 356–361.

8. Another completely different way that the contacts of a relay could fail was if dirt or an insect got trapped in the spacing between contacts. If a fly or a moth, for example, happened to be sitting on the make contact when the coil was energized, then it could be squashed and, after its smashed little body dried, the contacts would be covered with a very disgusting but quite effective insulator. To clean up such a disabled relay was called *debugging*, a term that has survived in the vocabulary of modern computer users trying to fix their faulty programs. This is not a joke—I heard it as a quite serious story in a lecture at the Naval Postgraduate School in 1982 from a legend in computer science, Rear Admiral Grace Hopper (1906–1992), a Yale PhD

mathematician who worked during the Second World War with Harvard's five ton, 800 cubic foot Mark I relay computer, which when operating was described as sounding like a "roomful of ladies knitting." To debug such a machine must have been an "interesting" job for someone; the successor to the Mark I—called, not surprisingly, the Mark II—had 13,000 relays. These were not fast machines; representing numbers in the form of $\pm p \cdot 10^n$ (with the decimal p given to ten significant digits and n varying from -15 to $+15$) the Mark II's add, multiply, and divide times were 0.2 seconds, 0.7 seconds, and 4.7 seconds, respectively. The "clicking" of the Mark machines, and of other relay computers, remained a signature characteristic for all computers in the minds of many, even long after vacuum-tube and solid-state electronics had eliminated such sounds. In his 1956 story "The Last Question," for example, science fiction writer Isaac Asimov has the huge computer named Multivac—specifically said to contain relays, even though "Multivac" is short for *multiple vacuum tubes*, "clicking." Years later he repeated this idea in "Starlight!", in which Asimov has a computer on board an advanced interstellar spaceship of the distant future "clicking busily." That story was published in the October 1962 issue of *Scientific American* as part of an ad campaign for an electronics company—which probably means that none of their *engineers* vetted the tale!

6

Boole, Shannon, and Probability

We shall consider relay circuits in which the only causes of errors are of . . . two types—failure of contacts that should be closed to be actually closed and of contacts that should be open to be actually open. . . . A relay [with these faults] will be called a crummy *relay.*

—Claude Shannon, in his 1956 paper, "Reliable Circuits Using Less Reliable Relays"

6.1 A COMMON MATHEMATICAL INTEREST

Boole and Shannon shared a deep interest in the mathematics of probability. Boole's interest was, of course, not related to the theory of computation—he was a century too early for that—while Shannon's mathematical theory of communication and information processing is replete with probabilistic analyses. There is, nevertheless, an important intersection between what the two men did, and that's what I'll show you in this chapter. I will not go very deeply at all into what either man did with the subject of probability, but rather my intent here is to simply give you a flavor of how they reasoned and of the sort of probabilistic problem that caught their attention. Once we have finished with Boole's problem, you'll see that it uses mathematics that will play a crucial role in answering Shannon's concern about "crummy" relays.

Even though he had a doctorate in mathematics, Shannon was, at heart, an electrical engineer who happened to be particularly good with equations. Sometimes this dual interest got him into trouble with pure mathematicians. In what has since become an infamous episode in the lore of information theory, the University of Illinois probability expert J. L. Doob wrote a review (in a 1949 issue of *Mathematical Reviews*) of Shannon's *Bell System Technical Journal* paper from the previous year, "A Mathematical Theory of Communication." Doob's otherwise professional commentary was marred by the prissy remark, "The discussion is suggestive throughout, rather than mathematical, and it is not always clear that the author's mathematical intentions are honorable." That unnecessary statement clearly bothered Shannon and, nearly forty years later, he said of it (in his 1987 *Omni* interview): "I didn't like [Doob's] review. He hadn't read the paper carefully. You can write mathematics line by line with each tiny inference indicated, or you can assume the reader understands what you are talking about. I was confident I was correct, not only in an intuitive way but in a rigorous way. I knew exactly what I was doing, and it all came out exactly right."

Doob outlived Shannon, and some modern writers have claimed that before he died in 2004 he came to regret his 1949 words. Perhaps so, but it took a long time for his change-of-heart because ten years later he was on the attack again, now with an even harsher voice. Writing a guest editorial in, of all places, the *Institute of Radio Engineers Transactions on Information Theory*, he asked (after complaining about the lack of "theoretical results"), "Can it be that the existence of a mathematical basis [to information theory] is irrelevant?" Talk about bringing the camel inside the tent and having it continue to aim in the wrong direction! In general, electrical engineers and applied mathematicians have, since Doob posed this question in March 1959, dismissed him as someone who simply did not understand the concerns of communication engineers. Modern engineers view Shannon's "Mathematical Theory" as his *Principia*, an achievement even greater than his switching theory use of Boolean algebra, and more than fifty years after Doob's sneer the *Transactions on Information Theory* is still in business.

6.2 SOME FUNDAMENTAL PROBABILITY CONCEPTS

This chapter will not make you a probability expert; indeed, the only mathematical background in probability required on your part is my assumption that the following everyday idea makes intuitive sense to you. Suppose we have a fair coin, where *fair* means that when we flip such a coin the probability the coin lands heads equals the probability it lands tails. (It isn't, to use gambling lingo, *loaded* to make one side more likely than the other.) Both probabilities are, of course, $\frac{1}{2}$, since any coin (fair or otherwise) must land either heads or tails (we are not allowing landing on edge!) and the sum of the face probabilities is 1. I believe that my dependence on intuition here would, somewhat paradoxically, be endorsed by the purist Doob. Indeed, in the same editorial in which he suggested that there is no need in information theory for a mathematical theory, he ended with "Can it be... that there is a context in which the word 'information' is accepted by general agreement and *used in an intuitive way* [my emphasis] and that no more is needed?"

Now, further suppose that we perform the repeatable experiment of flipping our fair coin twice in a row, an experiment we do over and over many times, observing what happens with each double-flip, and so are able to construct the relative frequencies of occurrence for each of the possible outcomes. For each double-flip, there are four possible outcomes:

First Flip	Second Flip
H	H
H	T
T	H
T	T

Mathematicians call this listing of all the possible outcomes the *sample space* of the experiment, with each of the four outcomes called a *sample point* in the sample space. If we imagine there is a probability associated with each sample point, then

$$\sum_{i=1}^{4} P(s_i) = 1. \tag{6.2.1}$$

In this notation, $P(s_i)$ is the probability of the sample point s_i, where $s_1 = (H, H)$, $s_2 = (H, T)$, $s_3 = (T, H)$, and $s_4 = (T, T)$. Equation (6.2.1) holds because we know that every time we perform the experiment, one of the four sample points is certain to occur. We just don't know which one.

We intuitively feel that each of the sample points in our double-flip sample space are equally likely[1]—this is sometimes given the technical name of the *principle of indifference*—and so $P(s_i) = \frac{1}{4}$ for all four i. This also fits nicely with our intuitive feeling that the probability of a sample point in the experiment is the probability of H (or T) on the first flip times the probability of H (or T) on the second flip, that is, $(\frac{1}{2})\ (\frac{1}{2}) = \frac{1}{4}$. The reason why we can multiply the individual flip probabilities together is because we are assuming that the individual flips are *independent*. That is, what happens on the first flip has no influence on the second flip.[2] And, of course, unless you believe in a sort of time travel, it is impossible for the second flip to influence the first flip!

We can take advantage of the idea of multiplying the probabilities of individual, independent flip outcomes to construct the mythical fair coin out of any real biased coin. I use the word *mythical* because it should be clear that to make a coin with a probability of showing heads that is *exactly* $\frac{1}{2}$ is a most unlikely possibility; rather, the probability of heads will be some unknown p (perhaps *close* to $\frac{1}{2}$, but not *precisely* $\frac{1}{2}$), and the probability of tails will be the equally unknown $1 - p$. Now, consider the following procedure. Flip the coin twice. It shows (H,H) with probability p^2, (T,T) with probability $(1 - p)^2$, and either (T,H) or (H,T) with equal probabilities of $p(1 - p)$. So, if (T,H) appears, call it "heads" and if (H,T) appears, call it "tails." If either (H,H) or (T,T) appears, ignore the result and flip twice again. Thus, no matter what p is, with the uninteresting exceptions of $p = 0$ and $p = 1$, we have a fair outcome, that is, *exactly equal* probabilities for heads and tails![3]

Here's a useful thing we can do with the sample points of a sample space: think of a collection or set of the points as defining an event that occurs when the experiment is performed, as it is one of the sample points in the defining set that is the one that occurs. The probability of an event is the sum of the probabilities of the individual sample points that define the event. For example, suppose we define the events A

and B on the sample space of our double-flip experiment as

$$A = \{\text{“at least one head (H) occurs”}\}$$

and

$$B = \{\text{“both heads occur”}\}.$$

Then,

$$P(A) = P(s_1) + P(s_2) + P(s_3) = \frac{3}{4}, \tag{6.2.2}$$

while

$$P(B) = P(s_1) = \frac{1}{4}. \tag{6.2.3}$$

Notice that

$$P(AB) = \frac{1}{4} \tag{6.2.4}$$

because there is just one sample point, s_1, that is common to both A and B. $P(AB)$ is *not* equal to $P(A)P(B) = \frac{3}{16}$ because events A and B are not independent.

Finally, the one remaining general concept we need before we look in particular at Boole and Shannon is what is called *conditional probability*. Suppose somebody performs the double-flip experiment behind closed doors and so we don't know what happened. Then, they open the doors, stick their head out, and tell us that event A happened. With that new, additional knowledge, what now is the probability that B also happened? It is *not* $\frac{1}{4}$, the a priori (“before the fact”) probability of event B *before* the experiment was performed, because our newly acquired knowledge tells us that sample point s_4 did not occur. What we are after now is the so-called a posteriori (“after the fact”) probability of event B conditioned on the given knowledge that event A did indeed occur. Mathematicians write this conditional probability as $P(B \mid A)$. Before I show you how to calculate the value of

$P(B \mid A)$, I hope that the reverse conditional probability $P(A \mid B)$, the conditional probability event A occurred given that event B occurred, is obvious to you: $P(A \mid B) = 1$ because if two heads occurred, then we know with certainty that at least one head occurred!

To find $P(B \mid A)$, let's first generalize a bit beyond our double-flip experiment. Imagine that when we perform *any* experiment, any one of N different, equally likely outcomes is possible. If one of n_a of those sample points is the one that actually occurs, then we say event A occurred. If one of n_b of those sample points is the one that actually occurs, then we say event B occurred. We assume n_{ab} sample points are common to A and B. That is, if the sample point that occurs is one of the n_{ab}, then we say that both A and B occurred. (If, in fact, $n_{ab} = 0$ then A and B are mutually exclusive, which means $P(AB) = 0$.) Now, from all this we can write

(a) $$P(AB) = \frac{n_{ab}}{N};$$

(b) $$P(A) = \frac{n_a}{N};$$

(c) $$P(B) = \frac{n_b}{N};$$

(d) $$P(A \mid B) = \frac{n_{ab}}{n_b};$$

(e) $$P(B \mid A) = \frac{n_{ab}}{n_a}.$$

The probability in (d) has n_b as the denominator because, given that event B occurred, we know that whatever sample point occurred it had to one of n_b. Similarly for (e).

The total number of sample points in A or B, together, is $n_a + n_b - n_{ab}$, where we have to subtract n_{ab} because the first two terms each include n_{ab} and so, together, they count n_{ab} twice. So,

$$P(A + B) = \frac{n_a + n_b - n_{ab}}{N} = \frac{n_a}{N} + \frac{n_b}{N} - \frac{n_{ab}}{N},$$

or

$$P(A + B) = P(A) + P(B) - P(AB). \tag{6.2.5}$$

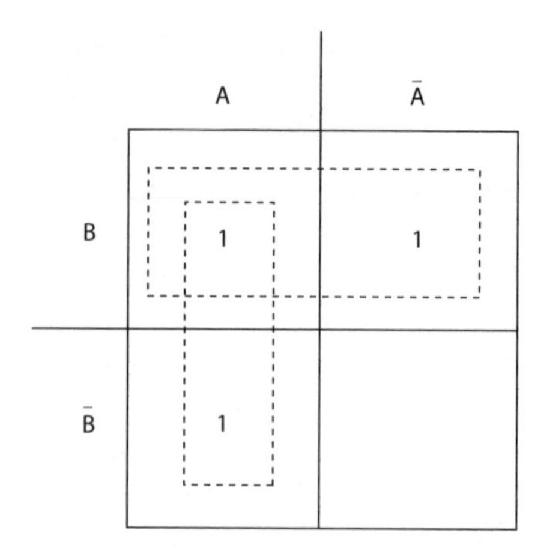

Figure 6.2.1. The map of $\mathbf{A}+\mathbf{B}$ covers $\mathbf{AB}$ *twice*.

Notice, carefully, that in (6.2.5) we are using the + symbol in two quite different ways: on the left it is the *logical* sum (the inclusive-or), while on the right it is the *arithmetic* sum. Also notice that only in the case of A and B being mutually exclusive is it true that $P(A+B)=P(A)+P(B)$.

We can interpret (6.2.5) in terms of a Karnaugh map, as shown in Figure 6.2.1, which shows the map for $A+B$. We now take each square covered as giving the probability of the square, and add the probabilities arithmetically (not logically); we subtract a probability that is covered more than once, which accounts for the third term in (6.2.5). When we get to Shannon's problem later in this chapter, you'll see how this probability interpretation of a Karnaugh map is very helpful.[4]

Okay, let's now calculate the conditional probabilities $P(B \mid A)$ and $P(A \mid B)$. From arithmetic we have

$$P(B \mid A)=\frac{n_{ab}}{n_a}=\frac{n_{ab}/N}{n_a/N}=\frac{P(AB)}{P(A)} \tag{6.2.6}$$

and

$$P(A \mid B)=\frac{n_{ab}}{n_b}=\frac{n_{ab}/N}{n_b/N}=\frac{P(AB)}{P(B)}. \tag{6.2.7}$$

So, putting (6.2.2) and (6.2.4) into (6.2.6), we have, for the double-flip experiment,

$$P(B \mid A) = \frac{1/4}{3/4} = \frac{1}{3},$$

which is significantly different from $P(B) = \frac{1}{4}$, the probability of B before the experiment is performed. And putting (6.2.3) and (6.2.4) into (6.2.7), we have

$$P(A \mid B) = \frac{1/4}{1/4} = 1,$$

which agrees with what I argued before should, in fact, be obvious for the double-flip experiment.

Here's another formula that is, I hope, equally obvious:

$$P(A \mid B) + P(\bar{A} \mid B) = 1,$$

which simply says, "Given that B occurred, then A either did or did not occur."

Finally, let's derive one last result that is sufficiently useful that it has its own name. The *theorem of total probability* says

$$P(A) = P(A \mid B)P(B) + P(A \mid \bar{B})P(\bar{B}). \tag{6.2.8}$$

All we need to show this is simple arithmetic, the observation that the number of sample points in the event $\bar{B}$ is $N - n_b$, and the observation that the number of potential sample points that can result in event A given that event B has *not* occurred is $n_a - n_{ab}$. Then,

$$P(A \mid \bar{B}) = \frac{n_a - n_{ab}}{N - n_b}.$$

And, of course,

$$P(\bar{B}) = 1 - P(B) = 1 - \frac{n_b}{N} = \frac{N - n_b}{N}.$$

So, substituting our earlier results for $P(A \mid B)$ and for $P(B)$, and putting these last two results into the right-hand side of (6.2.8), we have

$$\frac{n_{ab}}{n_b} \cdot \frac{n_b}{N} + \frac{n_a - n_{ab}}{N - n_b} \cdot \frac{N - n_b}{N} = \frac{n_{ab}}{N} + \frac{n_a - n_{ab}}{N} = \frac{n_a}{N} = P(A),$$

the left-hand side of (6.2.8) and we are done.

And, of course, we could equally well write

$$P(B) = P(B \mid A)P(A) + P(B \mid \bar{A})P(\bar{A}). \qquad (6.2.9)$$

6.3 BOOLE AND CONDITIONAL PROBABILITY

In the preface to his 1866 book, *The Logic of Chance*, John Venn (whose Venn diagrams are ancestors to Karnaugh maps, as noted for Chapter 4) wrote, "Probability has been very much abandoned to mathematicians, who as mathematicians have generally been unwilling to treat it thoroughly." This rather surprising statement was soon softened with some qualifications and explanations by Venn, but Boole—then dead for two years—would almost certainly have agreed with at least the first part of Venn's assertion. Boole wrote a good deal in the 1850s on conditional probability—what he called "the probability of causes"—and in particular about how shockingly ignorant so many of his scientific colleagues were of that mathematical topic.

In particular, Boole commented on what he called the "general doctrine... of the day" concerning the following situation. Suppose we are given two events, X and Y, defined on the sample space of some experiment, and are told that $P(\bar{X} \mid Y) = p$. Boole said that the "general doctrine" was that then $P(\bar{Y} \mid X) = p$, too. Interest in this particular mathematical issue was high in the 1850s because it appeared in a *theological* debate: was the existence of multiple star systems due just to chance, or would such systems be so unlikely to form by chance that their observed existence "proved" the intervention of a "Creator"? (As I mentioned back in Chapter 3, the initial spark to this debate was a 1767 paper in the *Transactions of the Royal Society* by the *Reverend* John Michell.) Well, of course, this is not a book to learn anything about theology, and I'll not say anything more along that line, but I do think you should know what brought Boole into such a debate in the first place.[5] Now, what of the *math* that Boole discussed?

The "general doctrine" does have a sort of plausibility to it: "if Y then not X" when "reversed" could be thought to imply "if X then not Y." Boole argued that this is not so, using the ideas of the previous section, and showed that $P(\bar{Y} \mid X)$ is given by a considerably more involved expression than simply "p." What Boole did was not really original, as conditional probability had been studied a century before by the English philosopher and minister Thomas Bayes (1701–1761), whose work was published posthumously in 1764 in the *Philosophical Transactions of the Royal Society of London*, where it was then promptly forgotten for twenty years until the great French mathematician Pierre-Simon Laplace (1749–1827) endorsed Bayes's results. What Boole did, then, with the following analysis, was remind his readers what the Reverend Bayes had done a hundred years before.

From the previous section—see (6.2.6) and (6.2.7)—we have

$$P(\bar{Y} \mid X) = \frac{P(X\bar{Y})}{P(X)}.$$

But

$$P(X \mid \bar{Y}) = \frac{P(X\bar{Y})}{P(\bar{Y})}$$

and so

$$P(X\bar{Y}) = P(X \mid \bar{Y})P(\bar{Y})$$

and so

$$P(\bar{Y} \mid X) = \frac{P(X \mid \bar{Y})P(\bar{Y})}{P(X)} = \frac{P(X \mid \bar{Y})[1 - P(Y)]}{P(X)}. \tag{6.3.1}$$

Or, from (6.2.8), we rewrite the denominator of (6.3.1) to get

$$P(\bar{Y} \mid X) = \frac{P(X \mid \bar{Y})[1 - P(Y)]}{P(X \mid Y)P(Y) + P(X \mid \bar{Y})P(\bar{Y})}$$

$$= \frac{P(X \mid \bar{Y})[1 - P(Y)]}{P(Y)\left[1 - P(\bar{X} \mid Y)\right] + P(X \mid \bar{Y})[1 - P(Y)]},$$

or, at last, we arrive at Boole's result (in our modern notation):

$$P(\bar{Y} \mid X) = \frac{P(X \mid \bar{Y})[1 - P(Y)]}{(1 - p)P(Y) + P(X \mid \bar{Y})[1 - P(Y)]} \tag{6.3.2}$$

which is (*most definitely*!) not just "p."

Now, finally, I can't end this section on the connection between Boole and probability without saying a few words about *Boole's inequality* (even if it's not *conditional* probability). I am not sure where in his writings Boole shows this result (or even if he ever actually did), but nonetheless his name has become attached to it. It is quite straightforward, and often quite useful, too. If $E_1, E_2, \ldots, E_k$ are *any* (not necessarily independent) k events defined on the same sample space, then

$$P(E_1 + E_2 + \cdots + E_k) \leq P(E_1) + P(E_2) + \cdots + P(E_k). \tag{6.3.3}$$

We saw a special case of (6.3.3) in (6.2.5). With the sample point concept in mind, (6.3.3) becomes what engineers and applied mathematicians call "trivially obvious" while textbooks written by pure mathematicians typically present a formal proof. Usually they use induction; (6.3.3) is an equality for the $k = 1$ case, and then if we assume it's true for $k = n$ it follows that it's true for $k = n + 1$. That's not really too awfully hard to understand, but even easier is the engineer's proof:

(1) The sample points in the superevent $E_1 + E_2 + \cdots + E_k$ are the sample points in E_1, plus the sample points in E_2 that haven't already been counted (remember, a sample point can be in more than one event), ..., plus the sample points in E_k that haven't already been counted; each included sample point has been counted exactly once.

(2) $P(E_1 + E_2 + \cdots + E_k)$ is the sum of the probabilities of all the sample points in (1), with the probability of each sample point appearing exactly once in the sum.

(3) Since a sample point can be in more than one event, then $P(E_1) + P(E_2) + \cdots + P(E_k)$ is the sum of the probabilities of all the sample points in each of the k events, with each sample point appearing *every time it occurs in any event*. Then (6.3.3) immediately follows, since probabilities are never negative.

6.4 SHANNON, CONDITIONAL PROBABILITY, AND RELAY RELIABILITY

Conditional probabilities abound in Shannon's 1948 "A Mathematical Theory of Communication," but what I'll show you here is their use in studying the reliability of relays. As discussed at the end of the previous chapter, relays are not the most reliable of electrical components. Shannon was particularly interested in how to reduce the probability of failure of a relay—see the chapter opening quotation—and what I'll do in this section is show you just the very surface of how he studied various schemes to do that.

Imagine that we have a supply of identical relays that each, with probability p, actuate when their coils are energized and successfully transition from their break contact to their make contact. Now, suppose that the value of p isn't good enough for some application, and we need to increase p. We could, of course, simply try to make a better relay, or we could (as did Shannon in 1956) try to see if it is possible to connect several of the original ("crummy," to use Shannon's own word)[6] relays together to arrive at an electrically equivalent circuit that has an increased effective value of p. For example, suppose we simply connect two of the crummy relays, A and B, so that their make contacts are in parallel, as shown in Figure 6.4.1 (the crosses represent the relays). The coils of A and B are wired in parallel so that the two relays are simultaneously energized; what then is the probability that there is an electrical path between terminals a and b?

This is an easy question to answer. There is no electrical path between a and b only if both A and B fail; that is, both $A = 0$ and $B = 0$.

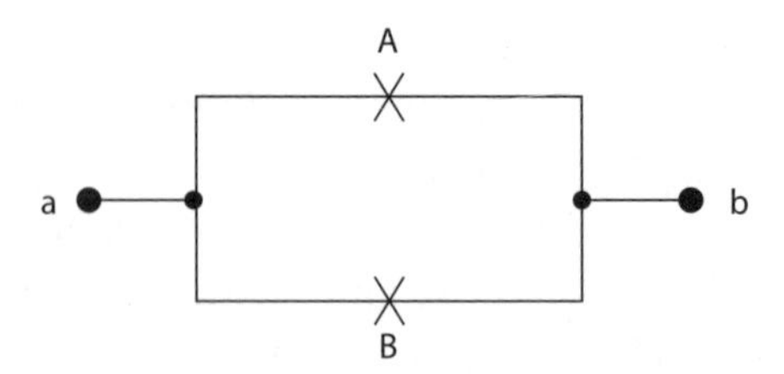

Figure 6.4.1. Two “crummy” relays with their make contacts in parallel.

(If, on the other hand, relay A, for example, works, then $A = 1$.) If the probability of a path through each relay is p, then the probability there is no path through A is $1 - p$, and similarly for B. If we assume A and B are independent, then both fail together with probability $(1-p)^2$, and the arrangement of Figure 6.4.1 fails with that probability. So, if $S1$ is the event that there *is* a path between a and b, then

$$P(S1) = 1 - (1-p)^2. \tag{6.4.1}$$

This is a useful thing to do if $P(S1) > p$. Is $p(S1) > p$? It is if

$$1 - (1-p)^2 > p$$

that is, if

$$1 - (1 - 2p + p^2) > p,$$

that is, if

$$2p - p^2 > p$$

that is, if (dividing by p, which means I am ignoring the possibility of $p = 0$, the uninteresting case of broken relays)

$$2 - p > 1,$$

that is, if

$$1 > p.$$

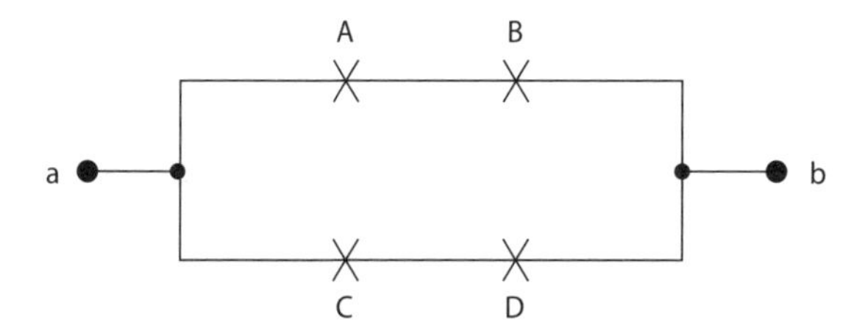

Figure 6.4.2. Four crummy relays in series/parallel.

And, of course, unless $p = 1$ (which means we started with perfect relays that never fail, and so the whole point of this analysis is moot) we always have $1 > p$, and so the connection in Figure 6.4.1 of two "crummy relays" is always more reliable than one crummy relay.

This result is probably not too surprising to you, but the arrangement of Figure 6.4.1 will serve as our standard against which we'll compare other more complicated arrangements of crummy relays, and so the result is worth demonstrating mathematically. Figure 6.4.2, for example, is a more complicated arrangement that shows a possible connection of four crummy relays that Shannon considered in his 1956 paper, and you should be able to see by inspection that the probability of the event $S2$ (there is an electrical path between terminals a and b) must be less, for all $0 < p < 1$, than $P(S1)$. That's because two crummy relays in series (either A and B, or C and D) is even crummier than a single relay (because $p^2 < p$ for $0 < p < 1$), and so Figure 6.4.2 is really just Figure 6.4.1 again, only now with relays having a smaller p than before. It will be useful to have an analytical expression for $P(S2)$ in just a bit, however, so I'll now work that out.

Each of the two parallel paths has a probability of working (both A and B work in the top path, or both C and D work in the bottom path) of p^2, and so each path has a probability of $1 - p^2$ of failing. For there not to be a path between a and b, both paths must fail, with aprobability of $(1 - p^2)^2$, and so

$$P(S2) = 1 - (1 - p^2)^2 = 2p^2 - p^4. \tag{6.4.2}$$

By plotting $P(S2) - P(S1)$ versus p I'll let you verify that the result is always negative, that is, $P(S2) < P(S1)$ for all $0 < p < 1$, as claimed.

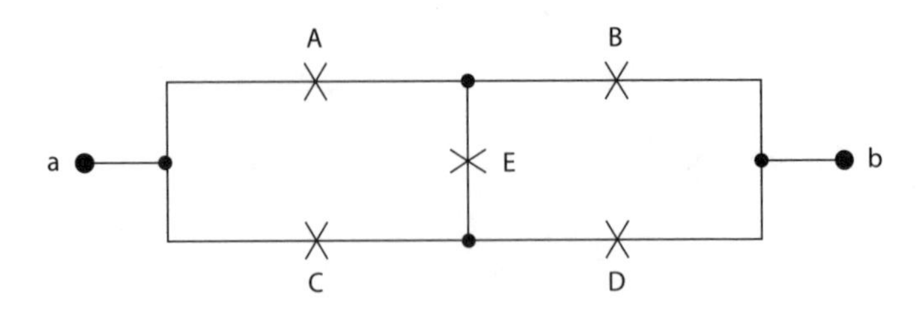

Figure 6.4.3. A bridging arrangement.

And now we come to our ultimate question in this section. Even though $P(S2) < P(S1)$ for all $0 < p < 1$, what if we now add a fifth crummy relay E as a bridge connection in Figure 6.4.2, to give Figure 6.4.3 (which we'll say has probability $P(S3)$ of working)? Does that extra relay path sufficiently increase the probability of an electrical path between a and b so that $P(S3) > P(S1)$? It isn't, I think, now quite so obvious as how to calculate $P(S3)$ as it was for the connection arrangements of Figures 6.4.1 and 6.4.2. Shannon gives the correct expression for $P(S3)$ in his 1956 paper, but he does not give its derivation. What I'll do now is show you a *possible* way to derive $P(S3)$ using conditional probability.

To start, fix your attention on the crummy bridging relay E, which either works ($E = 1$) or fails ($E = 0$). We can write, using the theorem of total probability—see (6.2.8)—$P(S3)$ as

$$P(S3) = P(S3 \mid E = 1)P(E = 1) + P(S3 \mid E = 0)P(E = 0). \qquad (6.4.3)$$

There are four probabilities on the right-hand side of (6.4.3), and we already know three of them:

$$\begin{aligned} &(a) P(E = 1) = p \\ &(b) P(E = 0) = 1 - p \\ &(c) P(S3 \mid E = 0) = 2p^2 - p^4. \end{aligned} \qquad (6.4.4)$$

The reason for (6.4.4c) is that if E fails then the bridge connection is not present and Figure 6.4.3 reduces to Figure 6.4.2, and so $P(S3 \mid E = 0) = P(S2)$. All we have left to do, then, is the calculation of

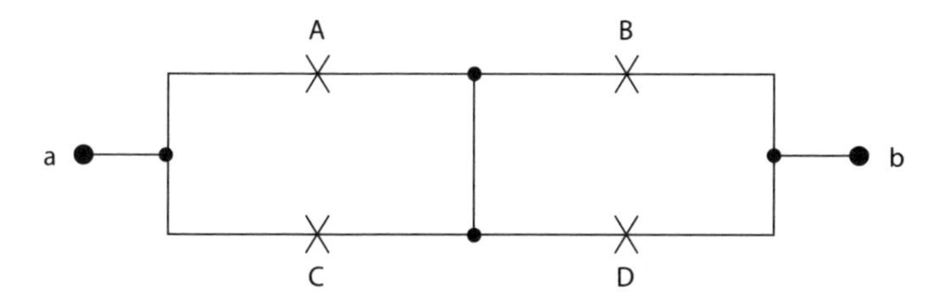

Figure 6.4.4. The bridge arrangement when **E** = 1.

$P(S3 \mid E = 1)$, which means we can redraw Figure 6.4.3 as Figure 6.4.4 in which relay E is replaced with a wire.

We can write directly from Figure 6.4.4 that

$$P(S3 \mid E = 1) = P(AB + CD + AD + CB). \qquad (6.4.5)$$

Notice, carefully, that we can *not* expand (6.4.5) as

$$P(S3 \mid E = 1) = P(AB) + P(CD) + P(AD) + P(CB)$$

because each of the four variables appears in more than one term, and so the individual terms are *not* independent. What we can do, however, is plot the Boolean expression $AB + CD + AD + CB$ on a four-variable Karnaugh map, as shown in Figure 6.4.5, and then use the probability interpretation of the map—take a look back at how we did this for (6.2.5)—to write $P(S3 \mid E = 1)$ as 1 *minus* the probability of the *inverse* map.[7] That is, as

$$P(S3 \mid E = 1) = 1 - P(\bar{A}\bar{C}) - P(\bar{B}\bar{D}) + P(\bar{A}\bar{B}\bar{C}\bar{D}). \qquad (6.4.6)$$

The fourth term on the right is there because the second and third terms, *each, subtract* the probability of the $\bar{A}\bar{B}\bar{C}\bar{D}$ map square and so we have to add that probability back in *once*. Now we can write directly from (6.4.6) that

$$P(S3 \mid E = 1) = 1 - (1 - p)^2 - (1 - p)^2 + (1 - p)^4$$

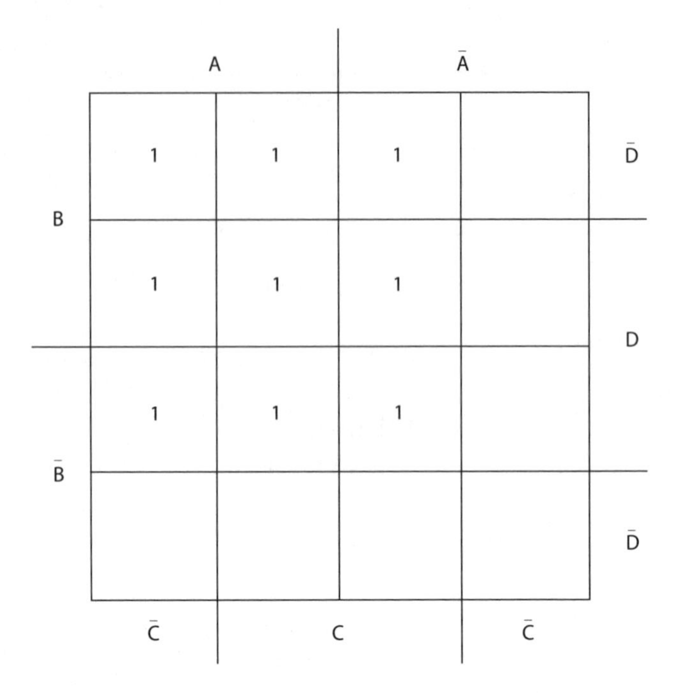

Figure 6.4.5. The Karnaugh map for $\mathbf{P}(S3 \mid \mathbf{E} = 1)$.

or, after just a little simplification,

$$P(S3 \mid E = 1) = 4p^2 - 4p^3 + p^4. \tag{6.4.7}$$

All the hard work is now done. We simply plug the probabilities from (6.4.4) and (6.4.7) into (6.4.3) to get

$$P(S3) = (4p^2 - 4p^3 + p^4)p + (2p^2 - p^4)(1 - p)$$

which reduces to

$$P(S3) = 2p^2 + 2p^3 - 5p^4 + 2p^5, \tag{6.4.8}$$

the result given by Shannon.

As I mentioned earlier, Shannon simply states (6.4.8) in his 1956 paper, without derivation, and so I don't really know if he used the conditional probability analysis I've shown you here. There is, in fact, another approach that he very well may have actually used, involving the inclusion-exclusion theorem (see note 4). As given in note 7, we have

$$P(S3) = P(AB + CD + AED + CEB).$$

If we define the four events E_1, E_2, E_3, and E_4 as $E_1 = AB$, $E_2 = CD$, $E_3 = AED$, and $E_4 = CEB$—notice that these events are not independent since A, B, C, and D each appear in multiple events, but that does not invalidate the theorem—then we have

$$\begin{aligned}
P(S3) &= P(E_1 + E_2 + E_3 + E_4) = P(E_1) + P(E_2) + P(E_3) + P(E_4) \\
&\quad -P(E_1E_2) - P(E_1E_3) - P(E_1E_4) - P(E_2E_3) \\
&\quad -P(E_2E_4) - P(E_3E_4) \\
&\quad +P(E_1E_2E_3) + P(E_1E_2E_4) + P(E_1E_3E_4) + P(E_2E_3E_4) \\
&\quad -P(E_1E_2E_3E_4) \\
&= P(AB) + P(CD) + P(AED) + P(CEB) \\
&\quad -P(ABCD) - P(ABED) - P(ABCE) - P(ACDE) \\
&\quad -P(BCDE) - P(ABCDE) \\
&\quad +P(ABCDE) + P(ABCDE) + P(ABCDE) \\
&\quad +P(ABCDE) - P(ABCDE) \\
&= p^2 + p^2 + p^3 + p^3 - p^4 - p^4 - p^4 - p^4 - p^4 - p^5 \\
&\quad +p^5 + p^5 + p^5 + p^5 - p^5
\end{aligned}$$

or,

$$P(S3) = 2p^2 + 2p^3 - 5p^4 + 2p^5$$

which is, again, just (6.4.8).

I didn't show you this derivation straightaway because I wanted first to show you some mathematics (conditional probability) used by Boole *and* by Shannon (in his *Mathematical Theory of Communication*). Also, I haven't actually derived, here, the inclusion-exclusion theorem for

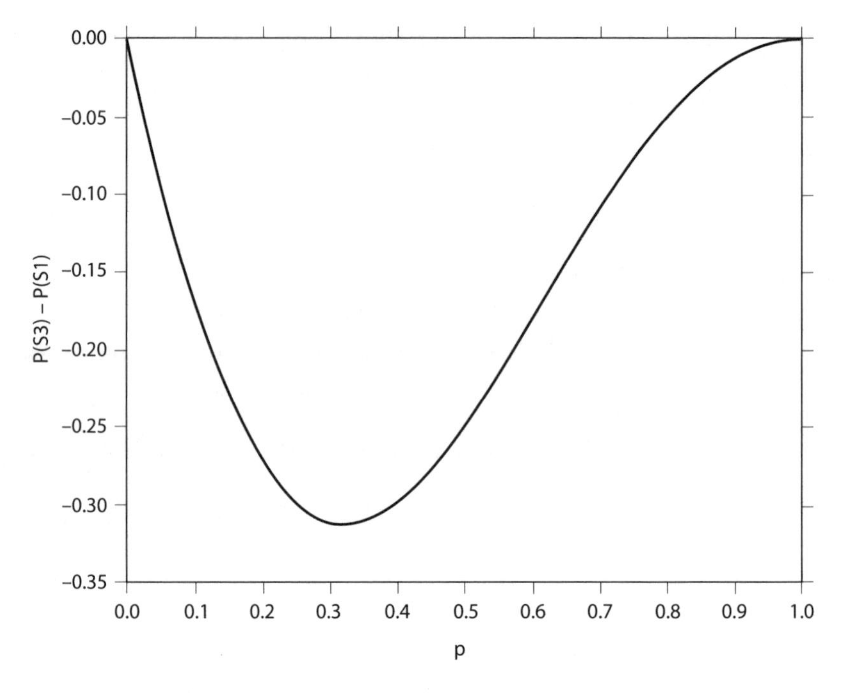

Figure 6.4.6. Relay **E** doesn't help.

the four-event case; in note 4 I left it up to you to fill in the (easy) details, while I did earlier derive the theorem of total probability and the conditional probability formulas that I used in the first derivation of (6.4.8).

In any case, however we get it, with (6.4.8) we can now answer our question of whether or not that additional bridging relay gives us a more reliable equivalent relay than do two relays in parallel. Figure 6.4.6 shows a plot of $P(S3) - P(S1)$ versus p, and we see that the result is always negative for $0 < p < 1$. So, the answer is that the additional relay E does *not* result in improved reliability.[8]

6.5 MAJORITY LOGIC

In this final section I'll comment just a bit on what sparked Shannon's interest in building more reliable circuits out of less reliable components. In 1956 Shannon was coeditor of an anthology of technical papers, one of which was authored by the great Hungarian-born American mathematician John von Neumann (1903–1957). Titled "Probabilistic Logics and the Synthesis of Reliable Organisms from Unreliable Components," Shannon read that paper as an editor long before the anthology appeared, and in his 1956 "crummy relay" paper specifically cited von Neumann's earlier work.[9]

Von Neumann's paper is heavily oriented toward mimicking the fundamental component of the human brain, the *neuron cell* that "fires" (that is, produces an output) when its multiple inputs (the outputs of other neurons) satisfy certain requirements. (A human brain has something like ten billion neurons, each connected to perhaps thousands of other neurons. There are, then, trillions of neural connections in a human brain.) Von Neumann was particularly interested in how a network of such cells could be constructed that would be able to self-repair damage to part of the network, or even reproduce itself from a large pool of "spare parts." One particular kind of very simple neuron of great interest to von Neumann was what he called a "majority organ," a cell with three inputs that fires when two or more (a majority) of its inputs are active. Von Neumann never discussed his majority organ in terms of specific hardware, but it is easy to build using relay technology—an observation surely not missed by Shannon.

You'll recall that a relay is energized when its coil current exceeds a certain critical minimum value, a value sufficient to produce a magnetic field large enough to physically move a spring-tensioned arm. Imagine now that we build our relay with three separate coils (each wound the same way so that each produces a magnetic field with north and south poles aligned), each coil with a resistance large enough so that when +5 volts is applied the coil current is only, say, two-thirds of the critical minimum required for energizing the relay. Thus, to energize the relay at least two of the three coils must receive +5 volts; the relay is energized only when a majority of its inputs are +5 volts. Alternatively, we could simply build a three-input majority circuit directly from NAND or NOR logic gates (built themselves from single-coil relays), and that should be an easy design task for you by now.

The use of such a majority gate has not (yet) caught on in the world of digital circuit design, although there is an active field of research in what are called neural nets (that is, networks of von Neumann's neuronlike "organs"),[10] and, if he had not died young, von Neumann would surely have been one of its most enthusiastic researchers. At the higher level of system design, however, the idea of a majority of inputs producing an output *has* caught on. To end this chapter on probability, then, let me show you an illustration of majority logic at the system level that checks for errors in the complete, computed solution to a problem that has been simultaneously solved by multiple machines.

My example of this can be found in the flight control computers of modern high-performance aircraft. Such aircraft are not "simple" things like a 1930s plane that had direct mechanical linkages from the pilot's cockpit controls (rudder pedal and control stick) to the aircraft's flight surfaces on the wings, tail, and rudder. The operation of the controls was fundamentally intuitive; for example, if you wanted to dive the airplane you simply pushed the stick forward. With today's huge commercial jumbo airliners flying close to the speed of sound, or a military combat jet flying at more than twice the speed of sound, asking the plane to dive and bank might involve quite complicated movements of the flight control surfaces (and the application of great force far beyond human limits to generate), which in turn would require nonintuitive physical motions of directly linked controls.

In modern aircraft the pilot's hand motions are still intuitive, but a digital flight control computer serves as a middle-man and takes the pilot's intentions from those hand motions and computes the proper signals to actuate powerful motors that move the control surfaces in such a way as to carry out those intentions. The computed solution process (called "fly-by-wire") is, of course, being continuously performed over and over, many times per second. If the flight control computer should fail, then the plane would literally become unflyable and we would have a virtually certain disaster.

To reduce the probability of such a disaster, suppose we install multiple flight-control computers, with each receiving the same input data, and so each solves the same control problem. These parallel solutions are done *independently* (to the point of having each computer with its own power supply and locating each computer far away from all the others). We then demand that the solutions from these independent computers agree. If one of the solutions suddenly starts

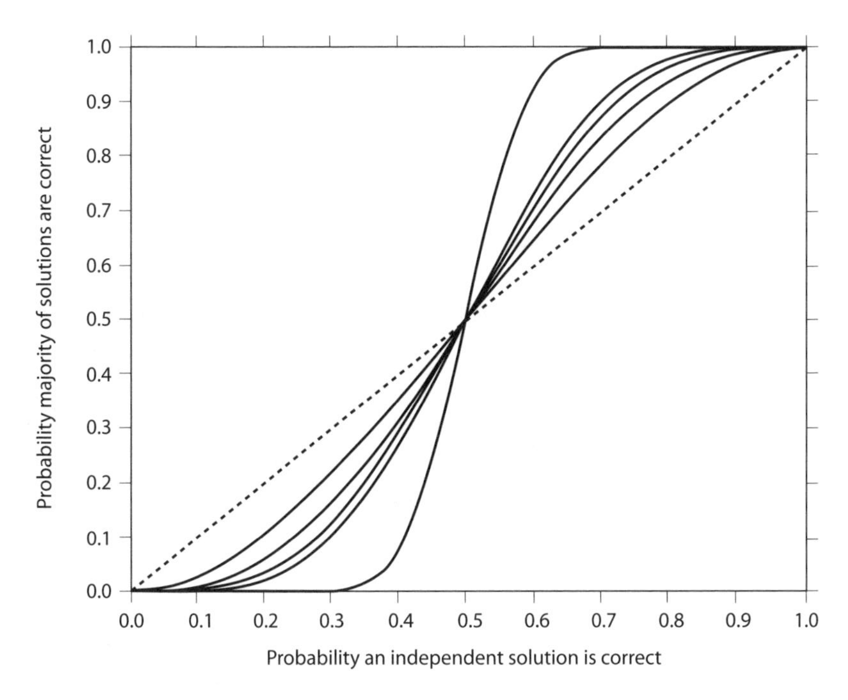

Figure 6.5.1. Majority logic for improved system performance.

to disagree with the others, we'll assume that solution is a rogue solution and simply ignore that solution's computer from then on. The aircraft will use only the common solution from the remaining computers that agree. We can continue to follow this procedure as long as we have a majority of the solutions in agreement. For example, if we initially have three independent computers, we'll have a correct solution as long as any two (or all three) solutions agree. If we start with n independent computers, with n odd, we'll have a correct solution as long as at least $\frac{n+1}{2}$ solutions agree.

So, here's our question: if p is the probability that a particular solution is correct,[11] what is the probability P the majority of n independent solutions (n odd) are correct? The theoretical answer is

$$P(n, p) = \sum_{k=\frac{n+1}{2}}^{n} \binom{n}{k} p^k (1-p)^{n-k}, \tag{6.5.1}$$

where the notation $\binom{n}{k} = \frac{n!}{(n-k)!k!}$ is the number of ways to select k correct solutions from n solutions (the other $n-k$ solutions are incorrect).[12] Here's why. The probability of each one of the possible cases of k correct and $n-k$ incorrect solutions is $p^k(1-p)^{n-k}$, and so the total probability *for a given* k is $\binom{n}{k}p^k(1-p)^{n-k}$. Summing over all relevant k, as in (6.5.1), gives us the above formula for P.

Figure 6.5.1 shows $P(n, p)$ versus p, as p runs from 0 to 1, for several values of n (I'll tell you what they are, soon). The dashed line is for the $n = 1$ case, and is included simply for reference. For $p < 0.5$, the majority logic curves are all below the $n = 1$ dashed line, and so in that case majority logic actually makes things worse. Of course, $p < 0.5$ represents a pretty crummy individual flight control computer (see note 11 again)! Suppose, however, that $p > 0.5$. Majority logic then gives improved performance compared to $n = 1$. The curves proceed upward from the dashed line for the cases of $n = 3$, $n = 5$, $n = 7$, $n = 9$, and, finally (and unrealistically but just for fun), for $n = 51$. Whether or not the increase of P over p is worth the increased cost of hardware required to build both the additional flight control computers and the circuitry to compare the multiple solutions is a management decision, not an engineering one.

NOTES AND REFERENCES

1. My discussion of sample spaces makes the implicit assumption that they are finite in size. All the formulas derived in the text are correct even for infinite sample spaces, but the derivations have to become a bit more sophisticated than those given here. In addition, pure mathematicians are not at all happy with my talk of equally-likely sample points as the starting point in the development, making the objection that *equally-likely* assumes the concept of probability as the starting point to developing the concept of probability. *Circular argument*, they shout—with some validity, too, I think—but reread Section 6.1, with Shannon's comment on Doob's criticism.

2. Many gamblers believe that if one flips a coin and a run of either tails or heads occurs, then the coin becomes "due" to show the other face. In fact, given that the coin is fair, then even if 100 tails show in a row (a not very likely occurrence, although not an impossible one), then on the 101st flip the

probability of a head is still just $\frac{1}{2}$. To believe otherwise, that the coin has a "memory" and keeps track of its past behavior, is called the "gambler's fallacy" and it inevitably leads to bankruptcy.

3. The price we pay for this method of making a fair coin from an arbitrary biased coin is time, because we have to flip the biased coin multiple times. In my book *Digital Dice* (Princeton University Press, 2008, pp. 248–251) I show that, on average, you have to double-flip the biased coin $\frac{1}{2p(1-p)}$ times to get a single decision, that is, either a "heads" or a "tails." If $p = 0.3$, for example, then you'll have to perform (on average) $\frac{1}{2(0.3)(0.7)} = \frac{1}{0.42} = 2.38$ double-flips of the biased coin. So, to get 100 decisions you should expect to double-flip 238 times, a total of 476 individual flips, compared to the 100 flips that would be required by an actual fair coin.

4. Eq. (6.2.5) Eq.is actually the first, most elementary special case of a far more general result called the *inclusion-exclusion theorem of probability*. If E_1, E_2, ..., E_n are any n events defined on the same sample space, then

$$P(E_1 + E_2 + \cdots + E_n) = \sum_{i=1}^{n} P(E_i) - \sum_{i=1}^{n} \sum_{j=i+1}^{n} P(E_i E_j) + \sum_{i=1}^{n} \sum_{j=i+1}^{n} \sum_{k=j+1}^{n} P(E_i E_j E_k) \ldots$$

Notice, *carefully*, that the n events are not required to be independent. Eq. (6.2.5) is the $n = 2$ case (with $E_1 = A$ and $E_2 = B$), and the $n = 3$ and $n = 4$ cases follow easily (do it!) from the probability interpretation of the three and four variable Karnaugh maps. You can find more on this theorem, including a detailed example of its use, in my book *Digital Dice* (see the previous note), pp. 237–243.

5. For a modern discussion on how binary star systems can be created through a physical process that must be a common occurrence in the universe—and so no supernatural intervention is required—see my book *Number-Crunching*, Princeton University Press, 2011, pp. 184–195.

6. The original title of Shannon's 1956 paper was "Reliable Circuits Using Crummy Relays," but that was changed at the request of the Bell Labs Public Relations Department. Shannon's sense of humor was not to be denied, though, and he managed to keep *crummy* in the text. Speaking of Shannon's sense of humor, there is a very funny story that Shannon himself was fond of telling that further illuminates his own lack of self-importance. When once giving a talk at the Institute for Advanced Study at Princeton, Albert Einstein came into the room and stayed at the back. After listening to Shannon speak for a few minutes, he leaned over to whisper into the ear of a nearby man; the

man whispered a reply and then Einstein quickly left. After the talk Shannon hastened over to the man to ask what the Great Man had said. "He wanted to know," Shannon was told, "which way to the nearest men's room."

7. We could, of course, have written directly from Figure 6.4.3 that $P(S3) = P(AB + CD + AED + CEB)$, and then used a Karnaugh map to get an expression for $P(S3)$ that would let us directly write P(S3) in terms of p. This would completely avoid the need for a conditional probability analysis. But it would require us to use a complicated *five*-variable Karnaugh map, which I haven't discussed in this book.

8. Shannon analyzed several other interconnection schemes of crummy relay switches in his 1956 paper, and you can find some further discussion in my book *Duelling Idiots and Other Probability Puzzlers*. Princeton University Press, 2000 (corrected paperback, 2002), pp. 22–23.

9. *Automata Studies* (edited by C. E. Shannon and J. McCarthy,) Princeton University Press, 1956, pp. 329–378.

10. What might happen when a neural net becomes really large has sparked the imaginations of science fiction writers as well as those of computer scientists. Arthur C. Clarke made good use of the idea in his "Dial F for Frankenstein." Set in the then future of 1975 (the story originally appeared in a 1964 issue of *Playboy*), the world's first satellite network has at last connected together all of the telephone exchanges on the planet. As one character explains,"Until today [our telephone networks have] been largely independent, autonomous. But now we've suddenly multiplied the connecting links, the networks have all merged together, and we've reached criticality." When another character asks what that means, the answer is chilling: "For want of a better word—consciousness." Since Clarke wrote his tale, satellites *have* interconnected all of the world's telephone networks, and nothing awful has happened. At least, I don't *think* so. Yet.

11. Initially, all n flight control computers are assumed to produce correct solutions. The value of p is the probability an individual computer produces correct solutions during the entire duration of the flight, from start to finish. The value of $1 - p$, then, is the probability a computer fails before the end of the flight, and P is the probability that a majority of the n computers work correctly during the entire flight. You should not think of p in the same way as you do (for example) about the probability of heads occurring when flipping a coin. A flight control computer doesn't randomly switch back and forth between correct and incorrect solutions. It *continuously* produces correct solutions (that is, solutions that are part of the majority of solutions) one after the other, until (if) it fails. With its first incorrect solution, it is disconnected from the remaining computers and thereafter ignored when solutions are compared.

12. The notation $\binom{n}{k}$ is called a *binomial coefficient* because of its appearance in the binomial theorem:

$$(x+y)^n = \sum_{k=0}^{n} \binom{n}{k} x^{n-k} y^k.$$

The interpretation of $\binom{n}{k}$ as the number of different ways to select k correct solutions from n solutions is easy to establish. For the first correct solution we have n possibilities, for the second we have $n-1$ possibilities, for the third we have $n-2$ possibilities, ..., for the kth correct solution we have $n-(k-1)=n-k+1$ possibilities. So, the total number of possibilities *where the order of selection matters* is

$$\begin{aligned}&(n)(n-1)(n-2)\cdots(n-k+1)\\ &\quad= \frac{[(n)(n-1)(n-2)\cdots(n-k+1)][(n-k)(n-k-1)\cdots(2)(1)]}{[(n-k)(n-k-1)\cdots(2)(1)]}\end{aligned}$$

or, using factorial notation, $= \frac{n!}{(n-k)!}$. But we don't care, in our discussion in the text, about the *order* of the selection for the k correct solutions, only that there are k correct solutions. Since there are $(k)(k-1)\cdots(2)(1)=k!$ ways to order the k selections, then the total number of ways for selecting the k correct solutions from n solutions *where order of selection is irrelevant* is $\frac{n!}{(n-k)!k!} = \binom{n}{k}$. Binomial coefficients occur all through mathematics, physics, and engineering (for some specific examples in physics, see my *Mrs. Perkins's Electric Quilt*, Princeton University Press, 2009, pp. 261–298), and whole books have been written on them. One such book, which has been in my personal library for decades, was written by one of Shannon's colleagues and collaborators at Bell Labs, the mathematician John Riordan (1902–1988); *Combinatorial Identities*, John Wiley 1968. One particularly beautiful identity comes immediately from setting $x=y=1$ in the binomial theorem, giving

$$2^n = \sum_{k=0}^{n} \binom{n}{k}.$$

It is with this identity that Riordan's book *begins* (actually, on p. 4), and by the time you reach the end (on p. 249) matters have gotten considerably more involved.

7

Some Combinatorial Logic Examples

If a computer can find out that there is an error, why can it not find out where it is?

—Richard W. Hamming, in his book *Coding and Information Theory* (1980)

7.1 CHANNEL CAPACITY, SHANNON'S THEOREM, AND ERROR-DETECTION THEORY

The entire point of Shannon's 1948 "A Mathematical Theory of Communication" was to study the theoretical limits on the transmission of information from point A (the *source*) to point B (the *receiver*) through an intervening medium (the *channel*). The information (for example, a human voice signal from a microphone or the output signals from the buttons of a keyboard) is imagined first to be encoded in some manner before being sent through the channel. In "Mathematical Theory" Shannon considers two distinct types of channels: the so-called *continuous channel* that would carry, for example, a continuous signal like the human voice, and the so-called *discrete channel* that would carry, again for example, a keyboard's output in the form of a digital stream of bits. For the rest of this chapter I'll limit my discussion to this second case. In a perfect world the digital stream would arrive at the receiver exactly as it was sent, but in the real world the channel is noisy and so, occasionally, a bit will arrive in error. That is, now and then a transmitted 0 will arrive as a 1, and vice versa.

Before continuing, just a note about the word "bit" which we'll use a lot in this chapter. It first appeared in the technical literature in "Mathematical Theory," with Shannon crediting its coining to the mathematician (at Bell Labs and then later at Princeton University) John W. Tukey (1915–2000). It's a contraction of "**b**inary dig**it**," and we've already seen its use in this book in Chapter 5, in the discussion on the "one-bit bi-stable relay latch." When talking of digital systems working with 1s and 0s, this is what most people, most of the time, mean by the word. In "Mathematical Theory," however, Shannon presented a new interpretation when those 1s and 0s are used to transmit information through a channel.

Imagine that you are sending a stream of bits through a channel and that *every one of them is the same*. That is, you are sending either all 1s or all 0s. Suppose you do this at, say, 1,000,000 bits/second. Shannon argued that the *information* rate would nevertheless still be zero! That's because you always know, before you even receive it, what the next bit will be: the same as the *last* bit. Shannon argued that to transmit information there must be *uncertainty* in what each new bit will be. For the case of a discrete channel carrying just 1s and 0s, for example, define p = probability a received bit is a 0 and $1-p$ = probability a received bit is a 1. Then, after stating several intuitively plausible properties that any useful measure of information should possess, Shannon showed that the non-negative quantity

$$H = -[p \log_2(p) + (1-p)\log_2(1-p)] \tag{7.1.1}$$

has those properties. This expression is the 2-symbol special case of the more general n-symbol case that Shannon discusses:

$$H = -\sum_{i=1}^{n} p_i \log_2(p_i), \quad \sum_{i=1}^{n} p_i = 1, \quad p_i \geq 0. \tag{7.1.2}$$

The formal resemblance of (7.1.2) to the entropy function in statistical mechanics (thermodynamics) is why Shannon called H, measured in bits, the *information entropy* function.

Now, to the central point here. In (7.1.1) if $p=1$ (all 0s) or if $p=0$ (all 1s), then $H=0$, just as Shannon argued on intuitive grounds. The maximum H in (7.1.1) occurs if $p=\frac{1}{2}$, which also makes intuitive sense: maximum information occurs when maximum uncertainty exists. Notice that $H = 1$ *only* if $p = \frac{1}{2}$, otherwise $H < 1$. Receiving a **bi**nary dig**it** does not necessarily mean you have received a bit of information. For that reason there has been proposed, from time to time, to call the *information unit* of H the *shannon*, where 1 shannon $\leq$ 1 bit with numerical equality only if 1s and 0s are equally likely.

One of the central concepts in Shannon's "Mathematical Theory" is that there are two numbers, C and R, called the *channel capacity* and the *source rate*, respectively, with both measured in units of bits/second (but keep in mind my comments about the "shannon"). For discrete channels Shannon showed that a reasonable definition for the channel capacity is

$$C = \lim_{T \to \infty} \frac{\log_2 N(T)}{T} \text{bits/unit time}, \tag{7.1.3}$$

assuming, of course, that the limit actually does exist, where $N(T)$ is the total number of different possible sequences of symbols (dots, dashes, and spaces, for example, in a telegraph channel) that can be sent in the time interval T. In "Mathematical Theory," Shannon gives a brief specific example of how to compute C, and I'll show you here in more detail how to do it. What makes this doubly interesting for us is that Boole's approach for solving a so-called *difference equation* with symbolic algebra can be used. (You'll recall from Chapter 3 that it was for his symbolic algebra to solve differential and difference equations that Boole received the Royal Society of London's 1844 Royal Medal in mathematics.) I won't use Boole's symbolic algebra to analyze the difference equation we'll encounter—it will be so elementary that using Boole's algebra on it would be like killing a fly with a cannonball—but I do think it curious that we now have a second example (along with probability) of how the mathematical interests of Shannon and Boole ran along parallel paths. The math I'll use requires no familiarity with calculus, but only with high school algebra, as I promised at the start of the book.

Suppose all messages to be sent over a telegraph channel are constrained to using a three-symbol alphabet: a *dot* occupying 2 units of time (telegraph key closed for one unit then open for one unit), a *dash* occupying 3 units of time (key closed for two units then open for one unit), and a *space* occupying 4 units of time (key open for four units). In terms of bits, a dot is 10, a dash is 110, and a space is 0000. All possible messages of duration T are all the possible different ways of stringing these three symbols together to form sequences of duration T. In this problem it should be clear that only integer values of T are of interest. So, for example, $N(0) = 0$, $N(1) = 0$, $N(2) = 1$, and $N(3) = 1$ because there are no sequences of durations 0 and 1, and just one sequence each of durations 2 (a dot) and 3 (a dash). Can you see that $N(4) = 2$ and $N(5) = 2$, because there are two distinct symbol sequences of duration 4—(dot, dot) and (space), and there are two distinct symbol sequences of duration 5—(dot, dash) and (dash, dot). Also $N(6) = 4$ with the symbol sequences being (dot, dot, dot), (dot, space), (space, dot), and (dash, dash).

Now, make the following simple observation: for all messages of duration $T > 4$, every one of them ends either with a dot, or a dash, or a space. That's because every sequence has to end with *something*! The number of messages ending with a dot must be $N(T - 2)$ since only a dot can be added to a message of duration $T - 2$ to make a message of duration T. Similarly, the number of messages of duration T ending with a space must be $N(T - 4)$, and $N(T - 3)$ must be the number of such messages ending with a dash. Thus, we arrive at the difference equation

$$N(T) = N(T-2) + N(T-3) + N(T-4),\ T \geq 5. \tag{7.1.4}$$

Before I show you how to solve (7.1.4) and use the result to evaluate C in (7.1.3), let me first admit that we really don't need to use an analytical approach at all. We could, instead, simply use (7.1.4) *directly* to calculate N(T) and thus C. For example,

$$\begin{aligned} N(7) &= N(5) + N(4) + N(3) = 2 + 2 + 1 = 5 \\ N(8) &= N(6) + N(5) + N(4) = 4 + 2 + 2 = 8 \\ N(9) &= N(7) + N(6) + N(5) = 5 + 4 + 2 = 11 \end{aligned}$$

and so on. From these and additional values of $N(T)$ we can easily determine C in the limit as $T \to \infty$. This process is easy to do by hand

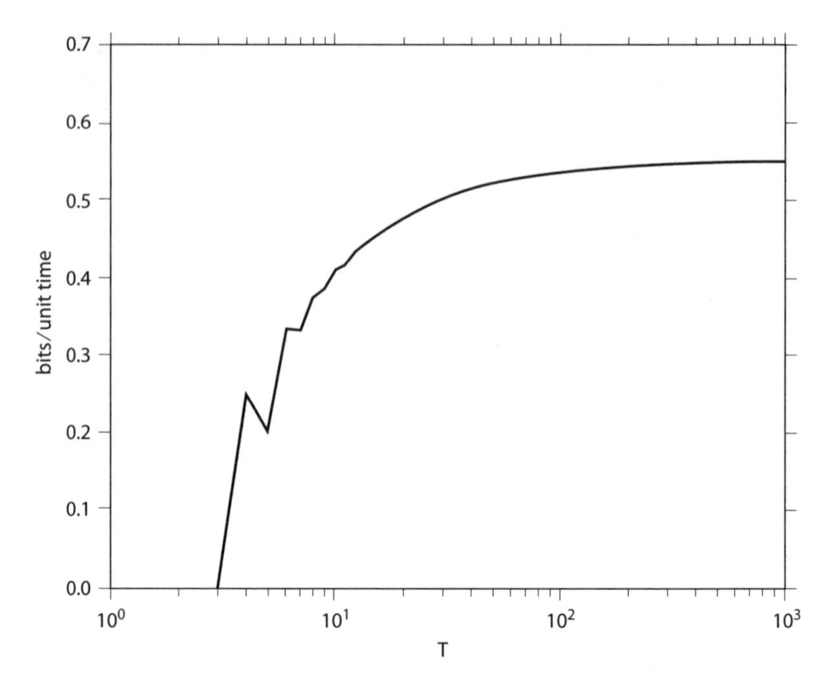

Figure 7.1.1. Channel capacity.

for *small* values of T, but a computer is much faster in general (the value of $N(T)$ increases very rapidly with increasing T; for example, $N(50) = 72,581,632$). Figure 7.1.1 shows a semilog plot of (7.1.3) over the interval $1 \leq T \leq 1{,}000$. As the plot suggests, it does appear as though the limit indeed exists, with a value of approximately $C = 0.55$ bits/unit time. If the unit of time (the time to send one bit) is 10 microseconds, then our telegraph channel has a capacity of about $C = 55,000$ bits/second.

We can solve (7.1.4) *exactly*—and thus both *prove* that the limit in (7.1.3) exists and calculate its precise value—as follows. Suppose we try (that is, *guess*—a perfectly respectable way to solve equations!) a solution to (7.1.4) of the form

$$N(T) = Ka^T, \tag{7.1.5}$$

where K and a are constants. We can confirm that this guess is a good one by simply substituting (7.1.5) into (7.1.4) and observing that it works. Doing that,

$$Ka^T = Ka^{(T-2)} + Ka^{(T-3)} + Ka^{(T-4)},$$

or, remembering how exponentials work, and canceling the K's (which means that the actual value of K isn't important), we have

$$a^T = a^T a^{-2} + a^T a^{-3} + a^T a^{-4}.$$

Or, since a^T cancels away (which means the specific value of T isn't important) we have

$$1 = a^{-2} + a^{-3} + a^{-4}$$

or, with a little rearrangement of terms,

$$a^4 - a^2 - a - 1 = 0. \qquad (7.1.6)$$

If you look at (7.1.6) for just a bit, you can see that one solution is $a = -1$. There are three other solutions, too, of course, and they are not so easy to see. Using either a computer (which is what I did) or the well-known formula for the solutions to a cubic (what (7.1.6) reduces to after removing the -1 solution), you'll find that all the solutions to (7.1.6) are

$$\begin{aligned} a_1 &= -1 \\ a_2 &= 1.46557\cdots \\ a_3 &= -0.23278\cdots + i0.79255\cdots \\ a_4 &= -0.23278\cdots - i0.79255\cdots, \end{aligned}$$

where $i = \sqrt{-1}$. Since each of these values gives a valid solution to (7.1.4)—as defined in (7.1.5)— the most general solution is the sum of them all (with each term having an arbitrary value of K). That is, the general solution to (7.1.4) is

$$\begin{aligned} N(T) = {} & K_1(-1)^T + K_2(1.46557)^T + K_3(-0.23278 + i0.79255)^T \\ & + K_4(-0.23278 - i0.79255)^T. \end{aligned} \qquad (7.1.7)$$

The four K's, as you'll see, do not have to be determined to compute C. That's because all we need in calculating C is an expression for $N(T)$ as $T \to \infty$, and in (7.1.7) there is just one term that grows without bound as $T \to \infty$, the second term. The first term simply oscillates between $\pm K_1$ as T takes on successive integer values. And since the complex conjugate solutions for a have an absolute value less than one, then the third and fourth terms in (7.1.7) actually go to zero as $T \to \infty$. So, in the *limit*,

$$
\begin{aligned}
C &= \lim_{T\to\infty} \frac{\log_2 N(T)}{T} = \lim_{T\to\infty} \frac{\log_2 \left\{K_2(1.46557)^T\right\}}{T} \\
&= \lim_{T\to\infty} \frac{\log_2 K_2}{T} + \lim_{T\to\infty} \frac{T \log_2(1.46557)}{T} \\
&= \log_2(1.46557) = 0.55146 \text{ bits/unit time,}
\end{aligned}
$$

in excellent agreement with Figure 7.1.1. Our derivation has also shown that the limit in (7.1.3) does actually exist.

Now, against all intuition, Shannon showed that if the source rate $R \leq C$ then there exists at least one source encoding procedure such that, *no matter what the channel noise may be*, the error rate at the receiver can be made arbitrarily small.[1] But, even with an arbitrarily small error rate it isn't zero, and thus there will still be errors. So, at the very least, the receiver would find it useful to be alerted when an error does occur (perhaps, if it's possible, to request a retransmission). That is, our first question here is: how can we encode the source information to allow *error detection* at the receiver? That can be done quite easily with what is called *parity*.

Imagine that when the source encoder is preparing the binary data stream for transmission through the channel, the encoder works with m information or *message* bits at a time. To these m bits the encoder appends one additional bit so that this new bit—called a *parity* bit—makes the total number of 1s even. We'll call these $m + 1$ bits a *block* . For example, if $m = 3$, then the message bits 011 would have a parity bit of 0 appended to give the 4-bit block 0110, while the message bits 100 would have a parity bit of 1 appended to give the 4-bit block 1001.

When the encoded $m + 1$ bit block arrives at the receiver, the m-received message bits are used to generate a new parity bit, which

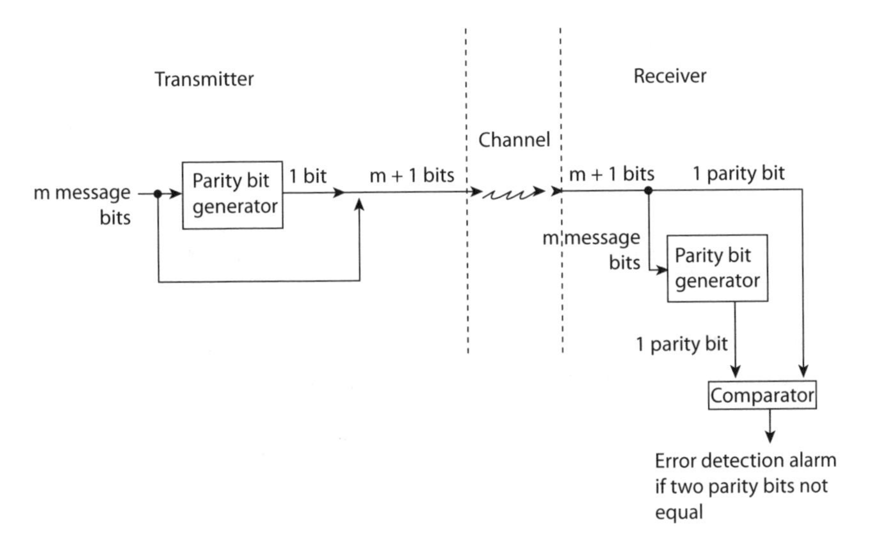

Figure 7.1.2. Error detection via parity.

is then compared to the received parity bit. If zero errors occurred in the message bits during transmission, then the two parity bits will obviously agree. If, however, *one* of the received message bits is in error (more generally, if an odd number of errors has occurred), then the two parity bits will not agree, and that disagreement can be used to trigger some sort of alert, as shown in Figure 7.1.2.

This approach to error detection is not without its problems, as there are two ways it can stumble. First, there may, in fact, be no errors at all in the received message bits, but the parity bit comparison at the receiver will still fail because it was the *transmitted parity bit itself* that was received incorrectly. And second, if there were actually two errors in the received bits (or more generally any even number of errors), then the parity bit comparison at the receiver will incorrectly say all is okay. A diagram like Figure 7.1.2 is very nice for a high-level, slide-show management meeting (I call it a *Jobs-diagram*, in honor of Apple's late marketing genius Steve Jobs, who sold a good line, but who I suspect might have been more than just a little vague on what is actually inside an Apple computer or iPad). For engineers who are tasked with building real hardware, however, it really won't do. What we need to do now is show *precisely* how to build both the parity bit generator logic at the source end of the channel, and the parity bit checking logic at the receiver end of the channel. What we are aiming for is a *Wozniak-diagram* (in honor of Apple's Steve Wozniak, the technical brains behind the original Apple computer).

7.2 THE EXCLUSIVE-OR GATE (XOR)

To lay the groundwork for parity logic, this section introduces a "new" logic gate, the exclusive-OR (written as XOR). It isn't actually really new, since it can be built from either NOR or NAND gates, but the XOR is such a useful function that it is usually treated as deserving a basic logic gate in its own right. If A and B are Boolean variables, then $f = A \oplus B$ is the exclusive-OR of A and B—take a look back at (4.3.5) —and its logic circuit symbol is shown in Figure 7.2.1. Unlike the other logic gates we've discussed, the XOR always has just two inputs. The construction of the XOR of A and B using just NOR gates is shown, and the circuit should make sense if you write (recall De Morgan)

$$f = A \oplus B = A\bar{B} + \bar{A}B = \overline{\bar{A} + B} + \overline{A + \bar{B}} = \overline{\overline{\overline{\bar{A} + B} + \overline{A + \bar{B}}}}.$$

How to build the XOR from AND, OR, and NOT gates should be obvious.

Before continuing with the parity discussion of the last section and how the XOR comes into play, let me digress for just a bit here and show you a fundamental application of the XOR in a completely different setting. The XOR is the basic circuit for what is called a *half-adder*, used in a computer's arithmetic circuitry. In binary arithmetic, 0 + 0, 0 + 1, and 1 + 0 are the same as in logic (that is, 0, 1, and 1, respectively). However, 1 + 1 = 1 in logic, but in binary arithmetic 1 + 1 is 0 *as well as* a carry-out of 1 to be added in the next most significant bit position. For example, in binary arithmetic we have

$$\begin{array}{ccc} 0 & 0 & 1 \\ 1 & 0 & 0 \\ \hline 1 & 0 & 1 \end{array}$$

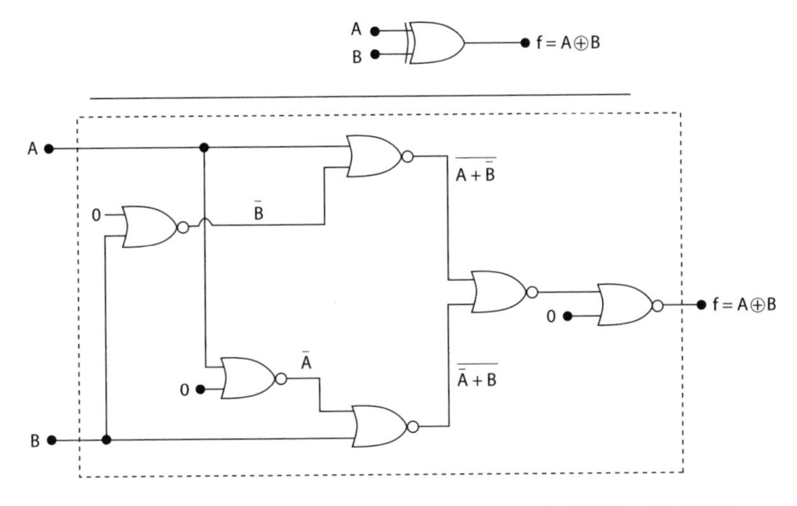

Figure 7.2.1. The XOR logic gate and how to build it with NORs.

(that is, 1 + 4 = 5 in decimal) in which no carry is produced in any of the bit positions, while

$$\begin{array}{r} 0\ 0\ 1 \\ 1\ 0\ 1 \\ \hline 1\ 1\ 0 \end{array}$$

(that is, 1 + 5 = 6) in which there is a carry produced in the addition of the two least significant (right-most) bits. And finally, consider

$$\begin{array}{r} 1\ 1\ 1 \\ 1\ 1\ 1 \\ \hline 1\ 1\ 1\ 0 \end{array}$$

(that is, 7 + 7 = 14) in which there is a carry produced in every bit position; notice that this also illustrates how adding two n-bit numbers can produce an $n+1$-bit sum because of a carry-out in the most significant (left-most) bit position.

To develop the logic circuitry to perform binary addition, let's start with the half-adder I mentioned earlier. A half-adder will generate as outputs the sum of its two input bits and a carry-out, but it does not

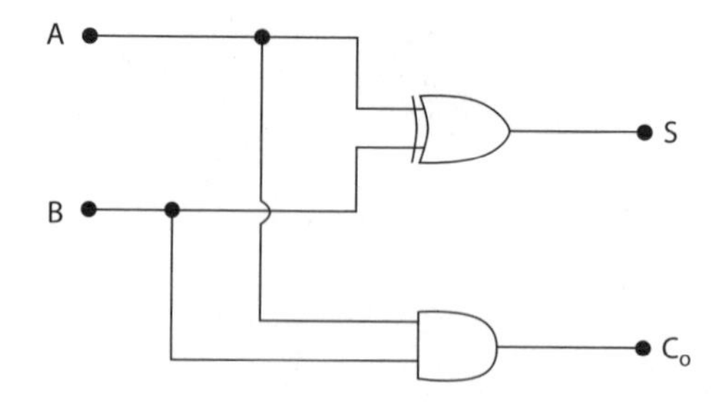

Figure 7.2.2. The half-adder.

accept a carry-out from the adjacent, lower significant bit position as a third input. A *full*-adder does accept a carry-out as a third input, and as you'll soon see we can make the obviously useful full-adder from two half-adders (which is why we start with the obviously limited half-adder!). The truth table for the half-adder, with inputs A and B and outputs S (for the sum) and C_o (the carry-out), is Thus,

A	B	S	C_o
0	0	0	0
0	1	1	0
1	0	1	0
1	1	0	1

$$S = \bar{A}B + A\bar{B} = A \oplus B, \tag{7.2.1}$$

and

$$C_o = AB. \tag{7.2.2}$$

The logic circuit for the half-adder, using one XOR and one AND, is shown in Figure 7.2.2.

The full-adder has the following truth table, where now, in addition to the inputs A and B, there is C_i, the carry-*in* which is the carry-*out* from the adjacent, lower significant bit position. The two outputs are, as before, S and C_o.

A	B	C_i	S	C_o
0	0	0	0	0
0	0	1	1	0
0	1	0	1	0
0	1	1	0	1
1	0	0	1	0
1	0	1	0	1
1	1	0	0	1
1	1	1	1	1

The carry-out equation is

$$C_o = \bar{A}BC_i + A\bar{B}C_i + AB\bar{C}_i + ABC_i = (\bar{A}B + A\bar{B})C_i + AB(\bar{C}_i + C_i),$$

or,

$$C_o = (A \oplus B)C_i + AB. \tag{7.2.3}$$

The equation for S is

$$S = \bar{A}\bar{B}C_i + \bar{A}B\bar{C}_i + A\bar{B}\bar{C}_i + ABC_i = (\bar{A}\bar{B} + AB)C_i + (\bar{A}B + A\bar{B})\bar{C}_i,$$

or,

$$S = (\bar{A}\bar{B} + AB)C_i + (A \oplus B)\bar{C}_i. \tag{7.2.4}$$

If you construct the Karnaugh maps for $\bar{A}\bar{B} + AB$ and $A \oplus B$, you'll see that they are inverses. That is,

$$\bar{A}\bar{B} + AB = \overline{A \oplus B},$$

and so (7.2.4) becomes

$$S = (\overline{A \oplus B})C_i + (A \oplus B)\bar{C}_i. \tag{7.2.5}$$

To make our next step transparent, write $W = A \oplus B$. Then we can write (7.2.5) as

$$S = \bar{W}C_i + W\bar{C}_i = W \oplus C_i,$$

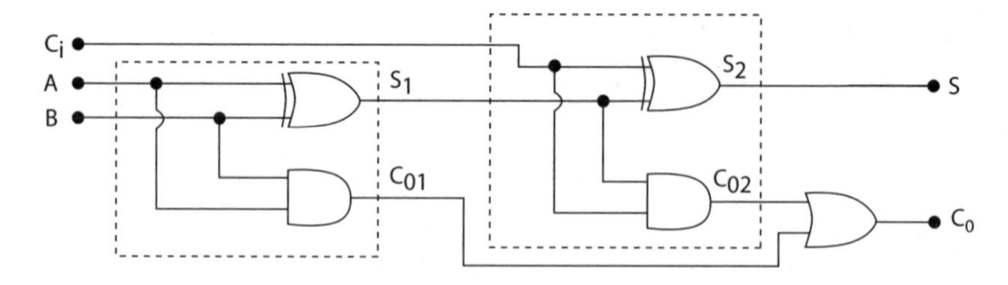

Figure 7.2.3. The full-adder (part 1).

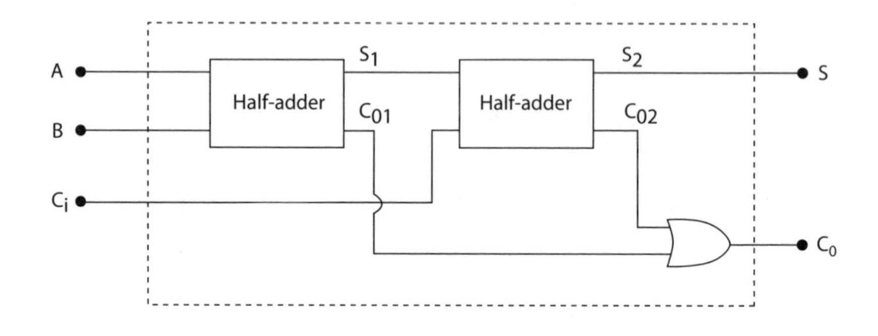

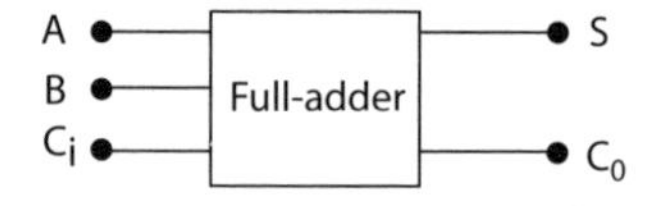

Figure 7.2.4. The full-adder (part 2).

or, finally,

$$S = (A \oplus B) \oplus C_i. \tag{7.2.6}$$

We can now draw, in Figure 7.2.3, the logic circuit for the full-adder directly from (7.2.3) and (7.2.6), using 2 XORs, 2 ANDs, and an OR. Each of the dashed boxes in that figure is a half-adder, as you can see if you look back at Figure 7.2.2. To make the two half-adder interconnections more obvious, take a look at Figure 7.2.4. We can now add (for example) two 4-bit numbers as shown in Figure 7.2.5, where in the least significant bit position the C_i input is hard wired to 0 (that is, is grounded) since, in that position, there is no carry-in from an

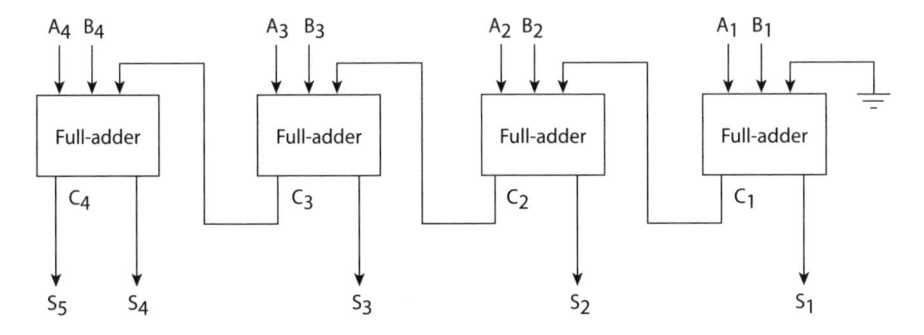

Figure 7.2.5. Adding two, 4-bit numbers.

earlier addition. To add numbers with even more bits, just use more full-adders.

Okay, now back to continue our error-detection discussion from Section 7.1 and how the XOR comes into play in that application.

7.3 ERROR-DETECTION LOGIC

There really is actually not much left to do, and this will be a (very!) short section. The XOR logic gate has an output of logical 1 only if its two inputs are different, and that is precisely what is needed to generate the parity bit from an odd number of 1 message bit inputs to give an even total of 1s in the transmitted block. That is, to generate the parity bit for any set of message bits, we simply XOR all the message bits as shown in Figure 7.3.1 for the case of three message bits (since the XOR gate always has just two inputs, we have to cascade XORs to handle three or more inputs). You should try various combinations of 0 and 1 inputs for m_1, m_2, and m_3 to convince yourself that the parity bit p will always be just what is required to give an even total number of 1s in the 4-bit block.

At the receiving end of the channel, we repeat the parity generator circuit using the received message bits as inputs, and then compare the newly generated parity bit with the received parity bit using one more XOR gate, as shown in Figure 7.3.2. If the two parity calculations disagree, we'll get an alarm. Notice that we'll get an alarm, even

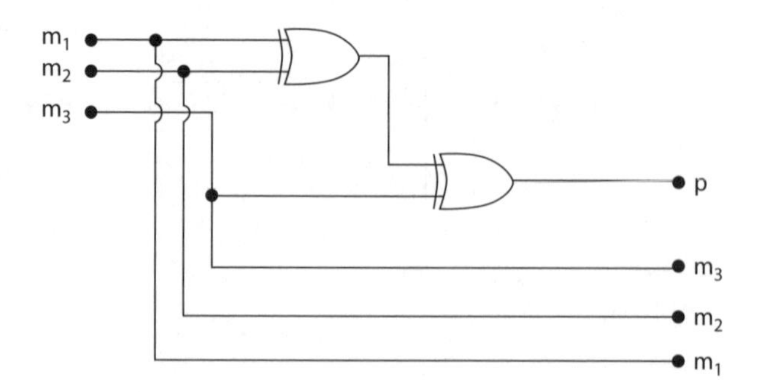

Figure 7.3.1. A parity-generator circuit (at source).

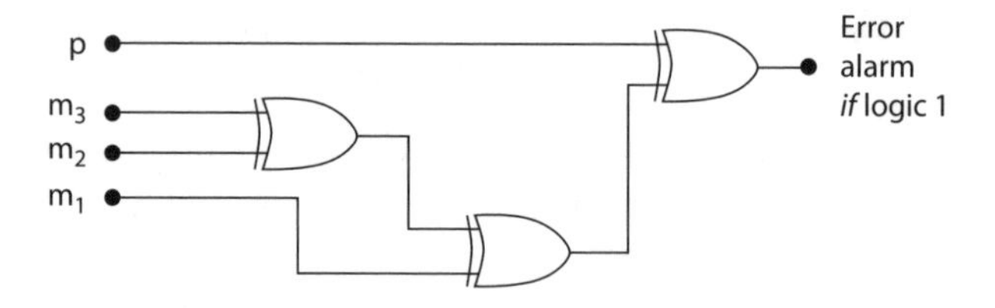

Figure 7.3.2. A parity-checking circuit (at receiver).

if the message bits are okay, if it is the parity bit that is received incorrectly.[2]

7.4 ERROR-CORRECTION THEORY

Generating an alert signal that indicates something is not quite right is good, but wouldn't it be even better if our digital logic could fix errors? Of course it would! Still, that seems at first glance to be as wonderful (and as fanciful) as it would be if you could levitate by simply pulling upward on your own feet. In fact, however, error correction *is* possible and, indeed, it's not even particularly hard to achieve. To see how to do it, suppose we have m message bits that we wish to send through a channel. To these m bits we'll append k additional bits, to arrive at a block of $n = m + k$ bits. (How these additional k bits are determined will be discussed in the next section.) There then are, of course, $2^n = 2^{m+k}$ possible blocks. Not all of these blocks, however, can represent an error-free transmission through the channel *if we wish to correct errors*. After all, if every possible block is associated with a legitimate message, how could we say any particular block is in error? What we need to do is to select a finite number—call it M—of the 2^n possible blocks and say just those particular M blocks (let's call them, from now on, *codewords*) are associated with legitimate messages. If anything other than one of the M codewords appears at the output of the channel, then we can say the transmission has been corrupted.

Now, one final definition before I pull the error-correcting rabbit out of the hat. Given two n-bit blocks, let's compare them bit by bit and call the number of times the comparison fails (that is, the number of bit positions that are different) the *distance* between the two blocks. The minimum distance is of course zero (the two blocks are identical), and the maximum distance is n (the two blocks differ in each and every bit position). This is often called the *Hamming distance*—after Shannon's colleague at Bell Labs, the mathematician Richard W. Hamming (1915–1998). We'll hear from Hamming again in the next section.

The central idea behind error correction is to pick the M codewords so that if no more than a given maximum number of errors occurs in an n-bit block—call that maximum t—then the distance d between any two codewords satisfies the inequality

$$d \geq 2t + 1. \tag{7.4.1}$$

Therefore, if no more than t errors occurs in a block during its journey through the channel, then the received corrupted version of the block will be closer to (that is, have a smaller Hamming distance from) the original codeword block than it will have from any other codeword. That means digital logic at the receiver can reset the corrupted block back to the original codeword sent by the source and, thus, the up to t errors will be corrected.

Achieving the required distance between codewords is the purpose of the k additional bits appended to the m message bits. There are 2^m different possibilities for the message bits, but 2^{m+k} possibilities from which to choose M of them to be codewords. If $k = 3$, for example, there are $2^3 = 8$ times as many possibilities from which to select the M codewords as there are 2^m message possibilities, and so the M codewords can be "spread sufficiently apart" from each other to achieve the required Hamming distance. Be quite careful to appreciate that this discussion does not tell us how to actually pick the special M codewords from the 2^{m+k} possibilities.

Before we discuss the selection of the M codewords, let's first calculate how big M can be. To do this, we need to first determine how many different ways a codeword can be corrupted into an "illegal" n-bit block by suffering exactly one error, exactly two errors, exactly three errors, up to exactly t errors. If we denote that value by $S(n, t)$, then

$$S(n,t) = \binom{n}{1} + \binom{n}{2} + \binom{n}{3} + \cdots + \binom{n}{t} \qquad (7.4.2)$$

where the notation $\binom{n}{k} = \frac{n!}{(n-k)!k!}$ is the binomial coefficient we encountered in the previous chapter.

We can now use a geometrical argument called *sphere-packing* to compute an upper bound on M. To do that, consider each n-bit block to be the coordinates of a point in n-dimensional space. That may sound very exotic, but let's start in our minds with the $n = 2$ and $n = 3$ cases. All 2-bit blocks are the four vertices of a square (a two-dimensional "cube"): (0,0), (0,1), (1,0), and (1,1). All 3-bit blocks are the vertices of a three-dimensional cube: (0,0,0), (0,0,1), (0,1,0), (0,1,1), (1,0,0), (1,0,1), (1,1,0), and (1,1,1). So, by analogy all 4-bit blocks are the vertices of a four-dimensional cube and, despite the science fiction sound of that, I'll bet you can now write down all sixteen of them. And so on for $n > 4$. Now, imagine that around each point associated with a codeword block we construct a surface of constant radius t; that is, a spherical surface that contains all the points that are no further than Hamming distance t from the center point (the codeword point) of the sphere. Since we have M codewords, we have M n-dimensional spheres.

If the channel delivers to the receiver an n-bit block that does not correspond to one of the M codewords, then the receiver looks to see which sphere the corrupted block belongs to (that is, which sphere the corrupted block's point is inside of) and then assumes that the codeword at the center of that sphere is what was actually sent by the source. To achieve error correction of up to t errors, we geometrically require that the spheres not intersect, a condition ensuring that for every received block there is a unique nearest codeword. Since there are $S(n,t)$ points inside each sphere plus one more (the codeword itself), and since there are M spheres, and since the total number of enclosed points can not exceed 2^n, we must then have $M[1+S(n,t)] \leq 2^n$ or, using (7.4.2),

$$M \leq \frac{2^n}{1+\binom{n}{1}+\binom{n}{2}+\binom{n}{3}+\cdots+\binom{n}{t}}. \tag{7.4.3}$$

For the simplest case of being able to correct a single error, we have $t=1$ and so

$$M \leq \frac{2^n}{1+n} = \frac{2^{m+k}}{1+m+k}.$$

For example, suppose $m=4$ message bits. Then

$$M \leq \frac{2^{k+4}}{k+5}.$$

If $k=2$, then $M \leq \frac{2^6}{7} = \frac{64}{7}$, or, since M is an integer, $M \leq 9$. But with $m=4$ message bits there are a total of 16 possible messages that could be sent and so we see that, with $k=2$, there simply aren't enough codewords separated widely enough to do the job. But, if we use $k=3$ then $M \leq \frac{2^7}{8} = \frac{128}{8} = 16$, equal to the number of possible messages. With $k=3$, then, it is not a priori impossible to imagine that those three additional bits appended in some way to our $m=4$ message bits might be sufficient to achieve single-error correction. Sufficient, that is, *if* we can find 16 different 7-bit blocks out of the 128 total of 7-bit blocks, such that each codeword has a Hamming distance of at least 3—see (7.4.1) again, with $t=1$—from all the other codewords.

Now, before I show you such a selection of 16 blocks—yes, that *is* possible to do, so you have not been the victim here of a pathetic shaggy dog story!—let's consider in just a bit more depth how the receiver logic might actually perform error correction. A quite straightforward way would be to have the M codewords permanently stored in the receiver, that is, the M codewords would be literally wired right into the receiver circuitry when the receiver is constructed. Then, every time an n-bit block comes through the channel, the receiver compares that block with each stored codeword and selects the codeword closest in Hamming distance to the received block. While that is certainly a possible approach, it is not the historical development of error correction. Rather, engineers have searched for codes in which there is sufficient internal structure that the received block itself—even if corrupted—contains all that is needed to *calculate* the nearest codeword. Such codes are called *systematic codes*.

7.5 ERROR-CORRECTION LOGIC

Probably the best known of the systematic codes are the *Hamming codes*, which Hamming published in 1950. Hamming had developed his codes years before, but the legal department at Bell Labs held up publication until the patent lawyers had finished their work. And so it must have been an interesting time in Hamming's private thoughts when the simplest of the Hamming error-correcting codes first appeared in print, not in Hamming's paper but two years earlier, in Shannon's "Mathematical Theory." Here's how Shannon described Hamming's code:

> There are two channel symbols, 0 and 1, and noise affects them in blocks of seven symbols. A block is either transmitted without error, or exactly one symbol of the seven is incorrect. ... An efficient code, allowing complete correction of errors ... is the following (found by a method due to R. Hamming): Let a block of seven symbols be $X_1, X_2, \ldots, X_7$. Of these X_3, X_5, X_6 and X_7 are message symbols and chosen arbitrarily by the source. The other three are redundant[3] and calculated as follows:
>
> X_4 is chosen to make $\alpha = X_4 + X_5 + X_6 + X_7$ even[4]
> X_2 is chosen to make $\beta = X_2 + X_3 + X_6 + X_7$ even
> X_1 is chosen to make $\gamma = X_1 + X_3 + X_5 + X_7$ even

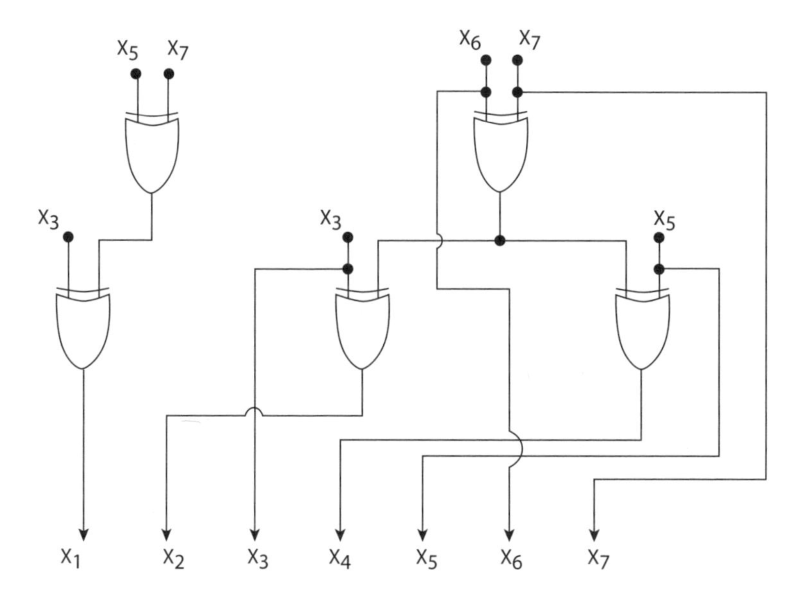

Figure 7.5.1. Single-error correcting Hamming encoder.

> When a block of seven is received α, β and γ are calculated and if even called zero, if odd called one. The binary number $\alpha\beta\gamma$ then gives the subscript of the X_i that is incorrect (if 0 there was no error).[5]

So, now we know what those k appended bits that we discussed in the previous section are; they are parity bits. Imagine that we write in binary the subscript i in X_i. Then, X_1 is the parity bit for the bit positions that have a 1 in the first position (least significant bit position). That is, X_1 is generated (at the source) from X_3, X_5, and X_7 (at the source, of course, we could equally well use m-notation for the message bits). X_2 is the parity bit for the bit positions that have a 1 in the second position; that is, X_2 is generated (at the source) from X_3, X_6, and X_7. And X_4 is the parity bit for the bit positions that have a 1 in the third position; that is, X_4 is generated (at the source) from X_5, X_6, and X_7.

To see how all this works, suppose the source wishes to send the four message bits X_3, X_5, X_6, and X_7 of 1011. The parity bits X_1, X_2, X_4 are calculated *at the source* as given above to be $X_1 = 0$, $X_2 = 1$, $X_4 = 0$,

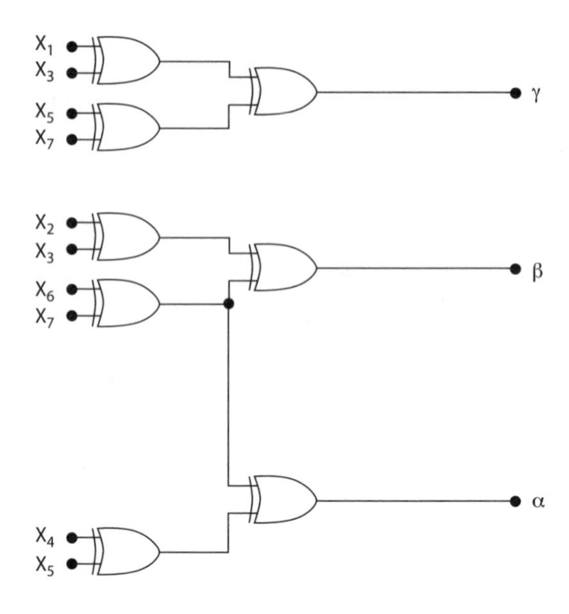

Figure 7.5.2. Syndrome generator (at receiver).

giving as the block to be transmitted 0110011. Now, imagine first that this block travels through the channel with no errors and arrives at the receiver just as sent. Then the binary number $\alpha\beta\gamma$ (called the *syndrome*, a word borrowed from the language of medical diagnosis) is calculated to be 000, the indication that indeed no error occurred. If, however, X_4 (for example) was received incorrectly, then the received block would be 0111011, and the calculated values of α, β, and γ would be $\alpha = 3$ (that is, odd, and so we set $\alpha = 1$), $\beta = 4$ (that is, even, and so we set $\beta = 0$), and $\gamma = 2$ (that is, even, and so we set $\gamma = 0$). Thus, the syndrome $\alpha\beta\gamma$ is calculated to be the binary number 100 which is 4, correctly indicating X_4 as the incorrect bit. Notice that this procedure can correct a single error in the parity bits just as easily as in the message bits (although, of course, correcting a parity bit once the block is through the channel is not of much practical interest).

The Hamming code described by Shannon corrects a single error in $m = 4$ message bits, but, in fact, if we are willing to append even more than $k = 3$ parity bits we can implement Hamming codes that can correct a single error in any number of message bits. Since the k parity bits can represent a total of 2^k binary numbers—a number that

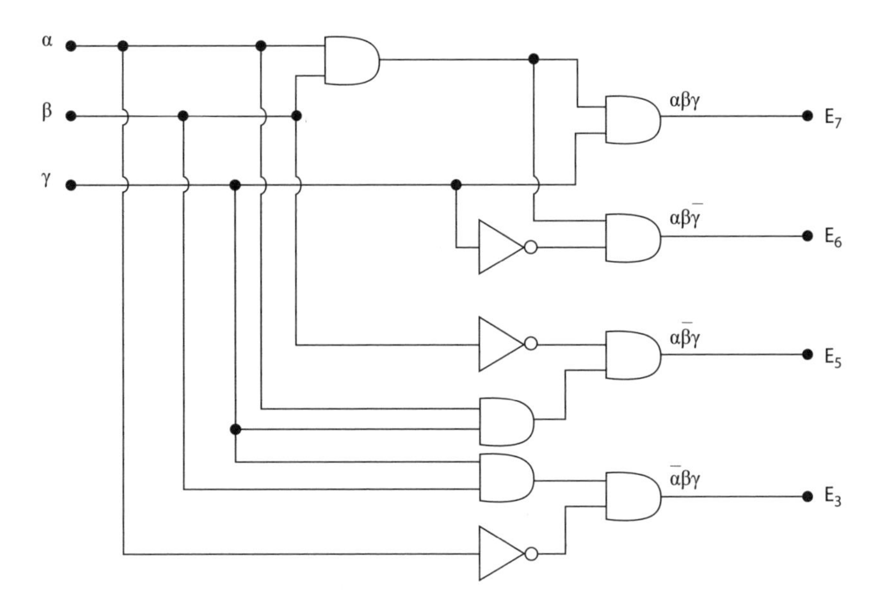

Figure 7.5.3. Syndrome decoder (at receiver).

has to be large enough to select one of the $m+k+1$ positions for the error location (the $+1$ term is for the "no error happened" indication, represented by the k-bit all-zeros binary number)—then we must have

$$2^k \geq m+k+1. \tag{7.5.1}$$

For the Hamming code in Shannon's paper, with $m=4$, (7.5.1) becomes the requirement $2^k \geq k+5$, which fails for $k=2$ ($2^2=4 \ngeq 2+5=7$), but which does work for $k=3$ ($2^3=8 \geq 3+5=8$). For $m=12$ message bits (for example), however, we'd need more than three parity bits. See if you can convince yourself that then $k=5$ parity bits are necessary for single-error correction.

Figure 7.5.1 shows how to build the logic circuitry that implements the generation (at the source) of the 7-bit block (4 message bits plus 3 parity bits) using single-error correcting Hamming encoding.

At the other end of the channel (at the receiver), Figure 7.5.2 shows the *syndrome generator* that accepts the received 7-bit block of X_i as inputs and generates as outputs the Boolean variables of α, β, and γ. The α, β, and γ outputs are the inputs to the *syndrome decoder*, whose

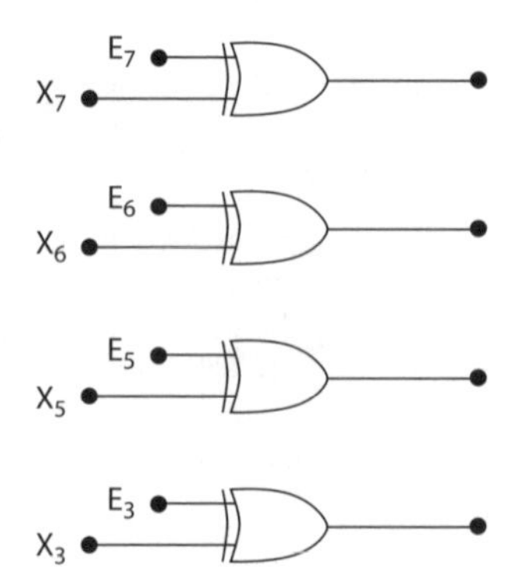

Figure 7.5.4. Corrected (if necessary) message bits (at receiver).

circuit is shown in Figure 7.5.3. The decoder has four output lines, one each only for the four interesting possibilities of the single error (if any) being in X_3, X_5, X_6, or X_7. So, if $\alpha\beta\gamma$ decodes as 3, 5, 6, or 7, then one of the output lines E_3, E_5, E_6, or E_7, respectively, is logic 1 and the other three lines are logic 0. If $\alpha\beta\gamma$ decodes as 0, 1, 2, or 4 (no error, or a parity bit error, and both cases mean the all-important *message* bits are okay), then all four output lines from the syndrome decoder are logic 0.

Figure 7.5.4 shows how the received message bits are processed to flip the bit (if any) that is in error. If none of the four output lines from the syndrome decoder are logic 1, then each of the four XOR gates in Figure 7.5.4 simply reproduces its input which is, of course, the correct thing to do since there was no error! If one of the decoder output lines *is* logic 1, however, then the corresponding XOR gate outputs the complement of its input,[6] which corrects an erroneous 0 back to 1, or an erroneous 1 back to 0.

NOTES AND REFERENCES

1. Shannon constructed his amazing proof *without* specifying any particular coding procedure. His ingenious demonstration went to the other extreme, in fact, in that he calculated the error rate averaged over *all possible* coding procedures. Since at least one member of any collection (or, to use the information-theoretic term, *ensemble*) being averaged must be *less* than the

average, then at least one of all possible ways to encode must have an error rate *less* than the average error rate. (A denial of this mathematical fact is, of course, the joke behind Garrison Keillor's claim—on his radio show *A Prairie Home Companion*—that the town of Lake Woebegone is "where all the children are above average.") Shannon's proof is what mathematicians call an *existence proof*. He left the detailed discovery of particular coding procedures with good (that is, small) error rates to others. In fact, Shannon created a cottage industry in code hunting! An older but still quite nice tutorial essay on this part of Shannon's work is by one of his Bell Labs colleagues, E. N. Gilbert, "Information Theory after 18 Years," *Science*, April 15, 1966, pp. 320–326. Coding procedures have, in the more than sixty years since Shannon's "Mathematical Theory" appeared, been developed that achieve error rates very close to his theoretical performance limits. There are many such codes in use today, going under such names as Reed-Solomon codes, convolutional codes, turbo codes, Hamming codes, BCH codes, ... These different codes are applicable in different situations. For example, one code is most useful when errors are rare and occur independently at random (caused, say, by a transient voltage spike), while another code may be capable of correcting what are called *burst errors* (*dependent* errors that occur one after the other because of, say, a scratch on a CD).

2. When I was a digital system designer back in the early 1960s, such a parity checking system was used on a machine I designed (for the data my machine received through a floor cable from another machine). All the parity error-alert signal did was turn on a push-button light on the control panel, which could then be turned off by pushing the button. If the light illuminated only infrequently, say once every half-hour or so (with a data transfer rate of a 22-bit block every 10 microseconds, that was a bit error rate of one error every four billion bits), we didn't worry about it. But, if the light came back on almost as soon as we had turned it off, then we knew something more serious was probably occurring (usually that someone had accidentally kicked the cable—because we weren't using a false floor—and loosened a connection).

3. The concept of redundancy is central in Shannon's "Mathematical Theory." Without redundancy you can't have error detection or correction (recall the text discussion in Section 7.4 on how, without the k-appended parity bits, all possible blocks would be associated with legitimate messages). Shannon's favorite example of redundancy was the English language, which he estimated to be 50% redundant (as he put it, "This means that when we write English half of what we write is determined by the structure of the language [that's the redundancy] and half is chosen freely.") It is redundancy that makes it possible for nearly anybody to correct the errors in the following sentence, which has been corrupted by deleting all the vowels:

MST PPL HV LTTL DFFCLTY N RDNG THS SNTNC

Redundancy is what makes it so easy for authors to miss typos when reading page proofs; the automatic error-correction of the eye/brain "computer" just reads right through them.

4. The + signs in the equations for α, β, and γ are arithmetic addition signs, and *not* inclusive-OR logic operations.

5. Claude E. Shannon and Warren Weaver, *The Mathematical Theory of Communication*, University of Illinois Press, 1949, p. 80. This book is simply Shannon's original paper plus an introductory expository essay by the mathematician Warren Weaver (1894–1978), then an administrator at the Rockefeller Foundation. The publisher apparently had so little confidence in Shannon as an author of clear prose that Weaver's essay was included to "clarify" Shannon. Weaver's contribution is questionable on two levels. First, it assumes the reader is the quite curiously odd person who both understands probability but still needs to be told (as Weaver actually does at one point) that the logarithm of a positive number less than one is negative! On an even more puzzling level, Weaver actually argues *against* Shannon's fundamental premise that the information content of a message has nothing to do with semantics. Weaver not only rejects that, he *deplores* it, and suggests ways to bring semantics back into consideration. Shannon must have ground his teeth at that when he read Weaver's essay. Weaver's conceptual roadblock survives today (see James Gleick's 2011 book *The Information*). It is, however, precisely because of Shannon's discarding of the romantic, emotional, false notion that messages somehow contain semantic meaning that is "beyond mathematics" that today we have codes that give us crystal-clear images from deep-space, and that let you watch streaming video on the Web at megabit/second speeds. Poets may decry this, but it isn't poetry that makes your e-mail possible; it's Shannon's "boring" (Gleick's word) *mathematical* information theory.

6. When used this way, the XOR is often called a controlled-NOT (CNOT) gate, and we'll see it and a more sophisticated version (a controlled-controlled-NOT) in Chapter 10.

8

Sequential-State Digital Circuits

0101100101100101100010

Every body continues in its state of rest, or of uniform motion in a right line, unless it is compelled to change that state by forces impressed upon it.

—From Isaac Newton's *Principia* (1687), showing that the idea of a physical system *changing state* long predates the invention of digital circuitry

8.1 TWO SEQUENTIAL-STATE PROBLEMS

What is a *sequential-state* problem? This is a question that is most directly answered by giving some specific examples. One *can*, I should admit, formulate a theoretical, mathematical definition, but examples are both more illuminating and, even more importantly, I think, more *fun*. My first example will drive home the point made by the opening quotation, that the concept of a physical system changing state with time predates Shannon and Boole by centuries. In fact, you'll see how the state concept predates *Newton* by even more centuries. This first example of the state concept comes from a late ninth century A D. manuscript of recreational math problems attributed to the English Catholic monk and scholar/educator Alcuin of York (735–804), who in 781 became head of Charlemagne's Palace School at Aachen, France.

Alcuin's math text—*Propositiones ad acuendos juvenes* (*Problems to Sharpen the Young*)—includes the following puzzle. On one side of a very wide river are two adults of equal weight, their two children who each weigh half as much as an adult, and one small boat. The boat is so small, in fact, that it can carry only the weight of one adult. Each of the four individuals can row the boat. How can the entire family get to the other side of the river?

We can answer Alcuin's question by defining a *state* to be a description of who (including the boat) is located where. To do this, I'll use the notation

state = [(who is on the starting side), (who is on the ending side)].

Writing **A**, **C**, and **B** to denote *adult*, *child*, and *boat*, respectively, we then have the given

initial state = [(**A**,**A**,**C**,**C**,**B**),()]

and the desired

final state = [(), (**A**,**A**,**C**,**C**,**B**)].

Notice that the empty parentheses () mean "nobody here". As the boat moves back and forth across the river, the state changes, with each crossing triggering a *transition* to the next state, all the while obeying the boat's weight constraint. The solution to Alcuin's puzzle, then, takes the form of a sequence of state transitions. Specifically, starting at the top with the initial state, the successive states are given by the following list:

[(**A**,**A**,**C**,**C**,**B**), ()]
[(**A**,**A**), (**C**,**C**,**B**)]
[(**A**,**A**,**C**,**B**), (**C**)]
[(**A**,**C**), (**A**,**C**,**B**)]
[(**A**,**C**,**C**,**B**), (**A**)]
[(**A**), (**A**,**C**,**C**,**B**)]
[(**A**,**C**,**B**), (**A**,**C**)]
[(**C**), (**A**,**A**,**C**,**B**)]
[(**C**,**C**,**B**), (**A**,**A**)]
[(), (**A**,**A**,**C**,**C**,**B**)].

We have a total (including the initial and final states) of ten states, with nine state transitions.

A variation of this puzzle has the two children wishing to finish being back on the original side of the river (with the boat, of course!), and the two adults on the other side. In this version the children are unrelated to the adults and, in fact, are simply running a transportation business. So, the initial state is as before, but now the final state is [(**C**,**C**,**B**), (**A**,**A**)]. The solution now is the original solution, plus one more state transition in which the two children return together to the original side of the river.

Notice that in Alcuin's puzzle we never revisit a state. The solution is a steady, unidirectional transition of states from the initial to the final. Many quite interesting problems do not have that property, and my second example illustrates that. In this second example I'll use a more modern situation that just about everyone of driving age has encountered: the automated parking garage. Suppose a garage has N parking spaces and two portals: an entrance and an exit, with each blocked by a movable guard arm. As you pull up to the entrance portal there is an electric sign that displays one of two messages: SPACE AVAILABLE, or FULL. If you see the second message you are out of luck and off you go. If you see the first message, however, you can pull up over a pressure pad that causes a conveniently placed dispenser to present you with a time-stamped card. Upon pulling the card out of the dispenser the blocking arm rises and you can then enter the garage to find an empty space (which you know must be *someplace* since the electric sign said so and, of course, electric signs never lie!).

The raising of the entrance portal blocking arm increments (by one) an automatic counter that was initialized to zero when the garage first opened. Similarly, when a car leaves the garage it approaches the exit portal (blocked by its arm), and the driver inserts the time-stamped card into an automatic reader, which then computes the bill. Once the driver inserts the proper amount of money into a conveniently located slot (or swipes a credit card), the blocking arm is raised and the automatic counter is decremented (by one). When (if) the counter reaches N, the FULL message is displayed. Otherwise, the SPACE AVAILABLE message is displayed.

The *state* of the garage is simply the number of cars in it (the number in the counter); the state can be any integer from 0 to N, and it changes, up or down, by 1 unless the garage is in state 0 (and then it can only go *up* by 1) or the garage is in state N (and then it can only go *down* by 1).

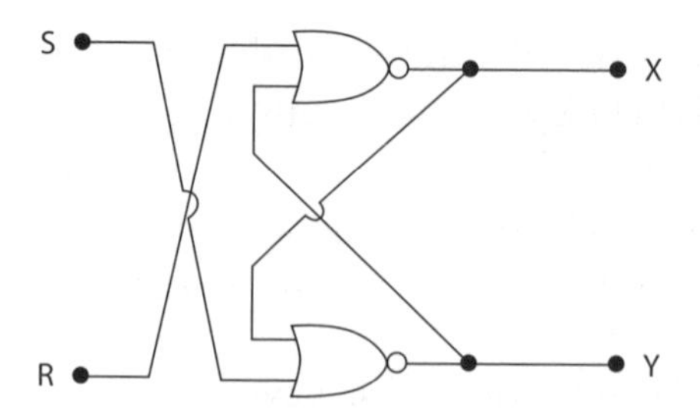

Figure 8.2.1. The NOR bi-stable latch.

The key element to building state-changing digital machines like our garage—or a modern digital computer—is the crude bi-stable latch that we first discussed back in Chapter 5. Since a latch has two stable states, then the totality of states that n latches can represent is at most 2^n. The design of a sequential state digital machine is simply that of determining how to interconnect these individual latches so as to transition from state to state with the goal of accomplishing a desired task. In the rest of this chapter I'll first show you how the latch can be modified in several important ways to arrive at the practical logic element called the *clocked, edge-triggered flip-flop*. Then we'll finish the chapter with the design of a specific machine, a machine that we'll see again in the next chapter when we discuss the famous Turing machine model for an arbitrarily powerful (but *not infinitely powerful*) computer.

8.2 THE NOR LATCH

A major problem with the bi-stable latch of Figure 5.6.1 is that its inputs are produced by push-buttons. What we'll need for useful sequential-state circuits is a latch with logic value inputs (+5 volts and ground for logic 1 and 0, respectively) that come from the outputs of *other latches and/or logic gates*. Such a latch is shown in Figure 8.2.1, made from two cross-coupled 2-input NOR gates. As in the relay latch of Figure 5.6.1, there are two inputs (S for *set* and R for *reset*) and two outputs (X and Y).

Here's how the NOR latch works. Let's start with $R = S = 0$. Then, the output of the upper NOR is

$$X = \overline{R + Y} = \overline{0 + Y} = \overline{Y},$$

and the output of the lower NOR is

$$Y = \overline{S + X} = \overline{0 + X} = \overline{X},$$

and these two results are obviously consistent. Thus, either $X = 0$ and $Y = 1$ (what I'll call state 1) or $X = 1$ and $Y = 0$ (what I'll call state 2). When the latch is first powered up, it will enter one of these two stable states, although which state is entered is indeterminate, depending on the relative speed of the particular components used to make each NOR. (When used in a real digital machine, latches should be initialized immediately after being powered up.) Let's now see what happens, starting with each state, as we change S and R.

Assume the latch is in state $1(X = 0, Y = 1)$ and that $R = 0$ as before. Then, let $S = 1$. So,

$$Y = \overline{S + X} = \overline{1 + 0} = \overline{1} = 0,$$

and

$$X = \overline{R + Y} = \overline{0 + 0} = \overline{0} = 1.$$

That is, the latch has transitioned to state 2. This is a *stable* transition, too, because if we do the calculations *again* with the new X and Y values, we get

$$Y = \overline{S + X} = \overline{1 + 1} = \overline{1} = 0$$

and

$$X = \overline{R + Y} = \overline{0 + 0} = \overline{0} = 1,$$

just as before. The latch *stays* in state 2, too, even when S returns to 0.

Notice that in these calculations I found Y first and then X. If we do the calculations in reverse order, we get

$$X = \overline{R+Y} = \overline{0+1} = \overline{1} = 0$$

$$Y = \overline{S+X} = \overline{1+0} = \overline{1} = 0,$$

and then do them again,

$$X = \overline{R+Y} = \overline{0+0} = \overline{0} = 1$$

$$Y = \overline{S+X} = \overline{1+1} = \overline{1} = 0,$$

and then do them yet again,

$$X = \overline{R+Y} = \overline{0+0} = \overline{0} = 1$$

$$Y = \overline{S+X} = \overline{1+1} = \overline{1} = 0,$$

just as we got when I calculated Y first and then X. That is, we see that $X = 1$, $Y = 0$ (that is, state 2) is a stable condition and, in the end, the order in which we calculate X and Y doesn't matter. Okay, now that we have the latch in state 2, we could ask what happens when $R = 1$ and $S = 0$. I'll let you repeat the above analysis, which, if you do, it correctly should lead you to the conclusion that the latch will transition back to state 1, and that the transition is a stable one even when R returns to 0.

What happens if we start in state 1 and then assume that both R and S are 1? Our equations tell us that

$$X = \overline{1+1} = \overline{1} = 0$$

$$Y = \overline{1+0} = \overline{1} = 0,$$

which is neither state 1 nor state 2. If we do the calculations again,

$$X = \overline{1+0} = \overline{1} = 0$$

$$Y = \overline{1+0} = \overline{1} = 0,$$

and so $X = 0$, $Y = 0$ as long as $R = S = 1$. We have lost the complementary relationship between X and Y. And matters don't get any better if we let S and R return to 0, because what happens depends on which logic signal returns to 0 first (and in any real-life machine it is certain that R and S will not be perfectly synchronized in time). To see this, suppose we first let $R = 0$. Then,

$$X = \overline{0+Y} = \overline{Y} = \overline{0} = 1,$$

and then, immediately after, we let $S = 0$ and so

$$Y = \overline{0+X} = \overline{X} = \overline{1} = 0,$$

which puts the latch in state 2. On the other hand, suppose we first let $S = 0$. Then,

$$Y = \overline{0+X} = \overline{X} = \overline{0} = 1,$$

and then, immediately after, we let $R = 0$ and so

$$X = \overline{0+Y} = \overline{Y} = \overline{1} = 0,$$

which puts the latch in state 1. The "solution" to this uncertain state transition situation is simplicity itself — we just avoid it! That is, we declare the $R = S = 1$ condition to be *forbidden*. (Just remember the old joke about how to make your head stop hurting because you've been repeatedly banging it into the wall —*stop banging your head into the wall*.)

Now, instead of drawing the detailed latch circuit diagram of Figure 8.2.1 over and over in our logic diagrams, I'll follow convention and use the simple logic symbol of Figure 8.2.2, with its inputs still as S and R, but (by convention) with its outputs as Q and $\overline{Q}$ (instead of X and Y, with the understanding that $X = \overline{Y}$). We'll call the ($Q = 1$, $\overline{Q} = 0$) state the *set* state, and the ($Q = 0$, $\overline{Q} = 1$) state the *reset* state. The input condition RS = 1 is forbidden, and it can occur only if the designer has "made a mistake." We call the end result the *RS latch*.

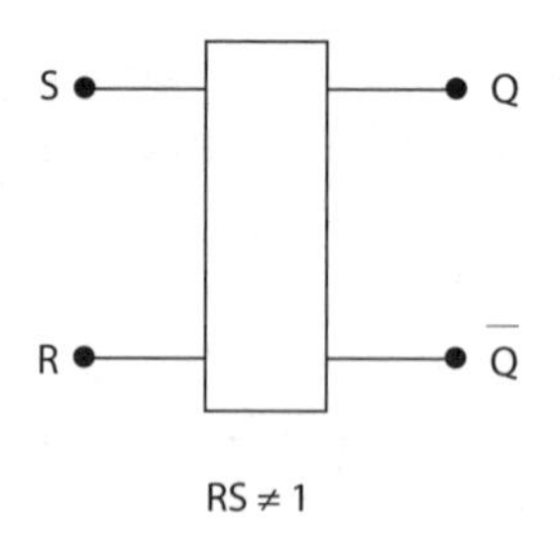

Figure 8.2.2. The RS latch.

8.3 THE CLOCKED RS FLIP-FLOP

The *RS* latch looks pretty simple, but it already can perform (before we soon embellish it with even more sophistication) some pretty neat tricks. Consider, for example, the following problem that a beginning digital circuit designer (your author, about fifty years ago) all too quickly thought easier to solve than it actually proved to be. Imagine that you have designed a digital machine that has a start push-button on its control panel. Until you push it, nothing happens. Push it, however, and your machine *goes*.[1] The "logical output" start signal produced by the button is to normally be 0 until you push the button, and then the logical output is to be 1 for a "brief" period of time (I'll say more about just what "brief" means, soon) and then return to 0. A very naive design to do this might be what I've drawn in Figure 8.3.1.

When PB is at its normal, lower position (on contact 1, which isn't connected to anything) the upper contact 2 is connected through the pull-up resistor r to +5 volts (logical 1), which means the logical output signal (the inverter output) is, as desired, logical 0. Then, when we push PB, contact 2 is directly connected to ground (logical 0), and so the logical output is 1 and it remains 1 until we release PB. (Without r, you should notice that when contact 2 is connected to ground, the +5 volt signal would be short-circuited—*not good*!) To achieve the "brief" period of time during which the logical output is 1, we'll simply punch the button! This indeed works—on paper but not in real life!

I learned that painful lesson on the very first logic design assignment I received after leaving graduate school in 1963. I remember a

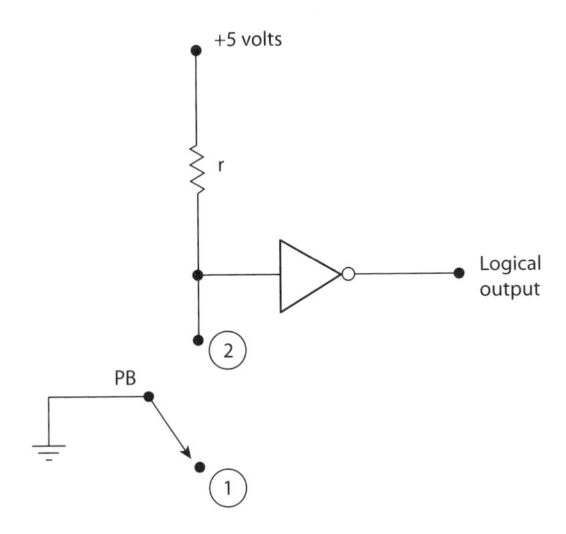

Figure 8.3.1. One way to generate the start signal for a digital machine.

more experienced colleague (of course, at that time *everybody* was more experienced than I was) patiently explaining to me that switches *bounce* on their contacts. That means that when PB is pushed to contact 2, it bounces on and off that contact several times over an interval of a few milliseconds before finally coming to permanent rest on contact 2. That means that rather than getting a *single* transition from 0 to 1 at the inverter output (logic designers call it a *clean* transition), we instead get *multiple* transitions from 0 to 1 to 0 to 1 to . . . to, finally, 1. This stutter-start signal might well cause problems in the circuitry that the start button is supposed to activate. And when we release PB, the switch will again bounce for a while, now off contact 1, although that generally doesn't cause any problems unless PB is such a crummy switch that it bounces all the way back to contact 2. A decent switch won't do that.[2]

One *non*logic way to eliminate switch bounce is to use a mercury switch. When the switch first hits contact 2 it lands in a pool of mercury; when bouncing occurs, surface tension pulls a filament of mercury up out of the pool, and so the electrical path is never broken. This solution works—but only if the switch never turns upside-down and is not subject to high vibration levels, events that often occur in cars driving on bumpy roads, in airplanes, in ships at sea, and on spacecraft.

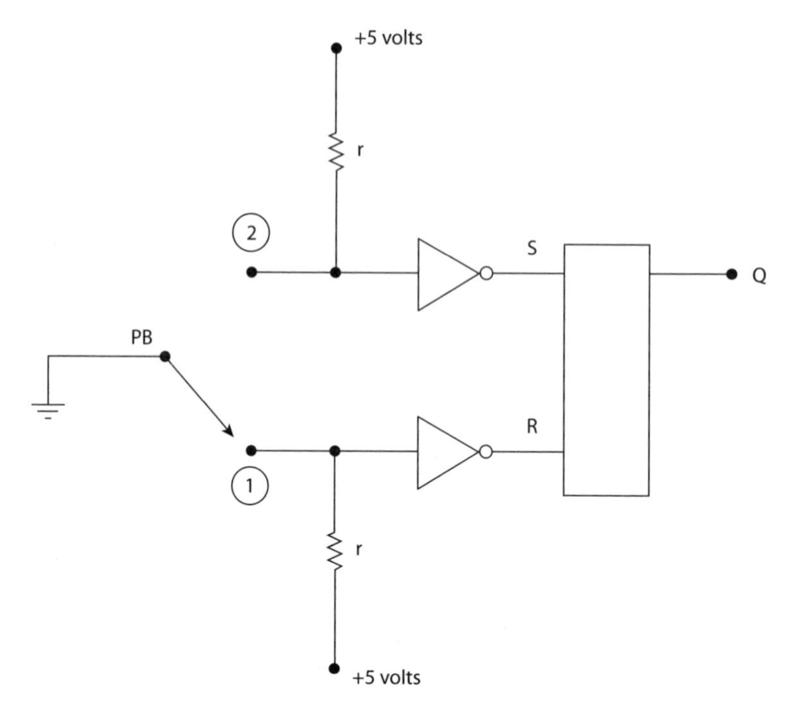

Figure 8.3.2. Switch debouncing with the RS latch.

Alternatively, and far better, we can logically eliminate switch bounce with the RS latch circuit of Figure 8.3.2. Called a *debouncer*, it works as follows. As before, PB is a push-button that is normally on contact 1, and so $S=0$ and $R=1$ and the latch is forced to be in the reset state ($Q=0$). Then, we push PB and the switch starts its journey to contact 2. At the instant the switch leaves contact 1, the input to the lower inverter is connected to $+5$ volts (through the lower pull-up resistor r) and so $R=0$. As soon as the switch arrives at contact 2 *for the first time* (remember, it will subsequently bounce for a bit), the input to the upper inverter is connected directly to ground (logic 0) and so $S=1$ (and $R=0$) and the latch *sets* ($Q=1$), *and nothing more happens after that until we release PB even though PB bounces on contact 2*. As the switch bounces, the value of S will alternate between 1 and 0, but, since the latch is already set, a fluctuating S is irrelevant. When we release PB, the switch moves back to contact 1, causing first $S=0$ and then $R = 1$ when the switch first hits contact 1, and so the latch resets ($Q = 0$). The switch bounces on contact 1, too, but that has no further influence on Q. We have a clean transition of Q from 0 to 1 when we push PB, Q stays at 1 until we release PB, and when we do release PB we have a clean transition of Q from 1 to 0. Notice that neither the physical orientation of the switch or vibrations will impact in the circuit of Figure 8.3.2.

Still, while the circuit of Figure 8.3.2 gives us a clean, bounce-free transition from 0 to 1, it is also a "long" duration signal. Even if you "punch" the push-button, the resulting start signal will spend a significant fraction of a second at logic 1. Our original description of the problem said we wanted a *brief* start signal and, to be specific, suppose we say it should be a pulse just *one microsecond* in duration. Nobody's finger is *that* fast! If you ask why so brief, let me remind you that when a race is started at a track meet it is with the single, *brief* report of a pistol shot, not by the sustained roar of a machine gun with the trigger held down. The start signal in a digital machine is used (among several tasks) to initialize the machine's starting state (that is, all latches are immediately either set or reset, as required), and we want to do that just *once*, not over and over as would happen with a "long" start signal. If this "explanation" leaves you still a bit unhappy—and perhaps that is actually a reasonable feeling at this point—then let me assure you that the solution to getting a one-microsecond pulse out of a mechanical push-button (you'll see how by the end of this section) will bring along with it the solution to another major problem in building digital machines that I haven't yet told you about.

This new, key idea is that of introducing what is called a *clock*. All modern digital machines I am aware of are sequential-state, *synchronous* machines. The "synchronous" adjective is there because all such machines require a periodic, pulselike timing signal (generated, usually, by a very stable, high-frequency crystal-controlled oscillator about whose electronic design we, as *digital* designers, need know nothing). Everything that happens in these machines is synchronized to this signal; in particular, latches can change state *only* at instants of time that are in step with the clock (a clock-synchronized latch is called a *flip-flop*). The clock signal is the heartbeat of a synchronous digital machine.

A metaphor for the clock that I would often use when teaching sequential-state design was that of the officer who beat the drum for the slave oarsmen on ancient Roman warships. His generic title was *pausarius*, Latin for "the timekeeper." It is easy to appreciate the chaos that would result if the rowers powering a ship became unsynchronized, and the same turmoil could similarly occur in a digital machine without a clock. Now, let me admit that a digital machine does *not actually have to be* synchronous, and in fact nonsynchronous (or *asynchronous*) machines have indeed been built. The theoretical appeal for asynchronous machines is that their circuitry doesn't have to repeatedly wait for the next clock pulse. Instead, *everything just happens as fast as possible*. The appeal of that potential speedup is undeniable, but the reality is less happy. Asynchronous machines have not met with much practical success, as they are extremely difficult to design, and even more temperamental in their operation. The sequential-state, clock-synchronized digital machine is going to be the realistic design model for a very long time to come.

Just because the state of a latch (that is, a flip-flop) can change only when a clock pulse occurs doesn't mean nothing is happening between clock pulses. In fact, the outputs of all the logic gates and latches in a synchronous digital machine are settling into their new values during that time interval, and then, at the next clock instant, those values are used to determine what the next state of the machine should be and the machine enters that state. And then the process repeats. Any particular logic signal may have to propagate both through several layers of logic gates before getting to one or more latches (with each gate having a very short *but not zero* delay, typically five nanoseconds or so—take a look back at Figure 7.2.1, for example, a circuit with four layers), as well as along various conduction paths (at the speed of light it takes one nanosecond to travel one foot, and electrical signals on wire travel slower than the speed of light). Once all these delays (determined by both the details of the electronic technology used and the logical complexity of the machine) are known, the minimum time interval between consecutive clock pulses (that is, the maximum clock frequency) can be calculated that allows enough time for all logic signals to stabilize. The actual clock frequency used is usually less than the maximum, to increase reliability, but some enthusiastic users (I'm

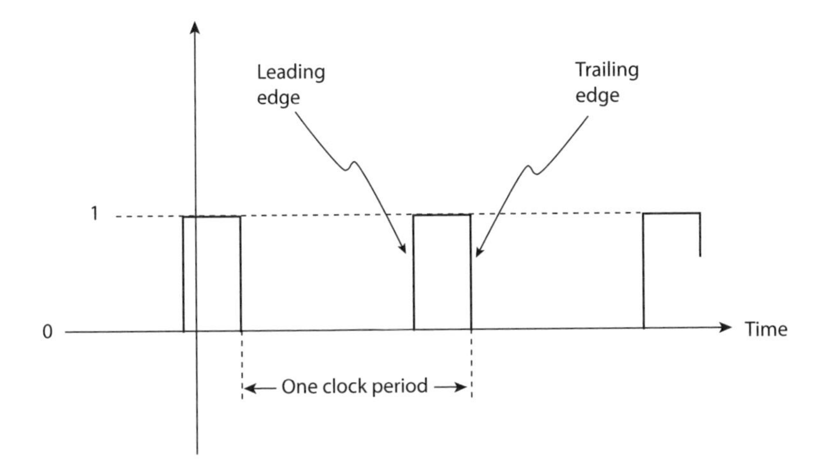

Figure 8.3.3. A clock signal.

thinking of video gamers) have been known to tweak the clock rates of their machines upward a bit to increase speed, a technique called *overclocking*.

A typical clock signal appears in Figure 8.3.3, which I've shown as periodically switching between logic 0 and logic 1. When the clock pulse transitions from 0 to 1, we call that instant the *leading edge* of the clock pulse, and the transition instant from 1 to 0 is called the *trailing edge*. For example, a 1 MHz (megahertz) clock would have a one-microsecond (1,000 nanoseconds) spacing between consecutive leading (trailing) edges, with (perhaps) a 20 to 50 nanosecond spacing between the leading and trailing edges of a given clock pulse. My 1963 machine (see note 1 again) had a clock frequency of 250 KHz (that's right, just $\frac{1}{4}$ MHz!). A fast commercial mainframe computer of that day probably had a 3 to 5 MHz clock. The 1970s CRAY-1, one of the early supercomputers, had an 80 MHz clock. By 1998, when I wrote my first book for Princeton, my everyday word processing laptop (an IBM ThinkPad 365ED) was nearly as fast, with a 75 MHz clock. The four-year-old (in 2011) PC I am typing this book on (a Dell Dimension 5150) has a 3 GHz clock (that is, 3,000 MHz). The fastest commercial mainframe computer in 2011 (the IBM zEnterprise 196) has a 5.2 GHz clock. That's nearly 21,000 times faster (more than *fourteen* doublings in speed!) than my 1963 machine.

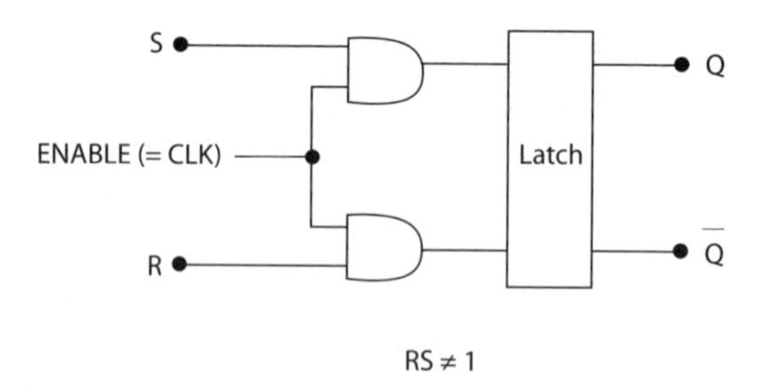

Figure 8.3.4. The RS flip-flop.

To synchronize an *RS* latch with the clock signal (CLK) to create an *RS* flip-flop, all we do is add two AND gates, as shown in Figure 8.3.4, with each AND gate having the signal ENABLE as an input. In the simplest case, ENABLE is CLK itself. Notice that in this circuit the inputs to the latch are present throughout the duration of the clock pulse and, as brief as that is, if *S* or *R* (or perhaps both) should change (for whatever reason) during the pulse duration, then the state of the latch might also change. To further minimize the likelihood of that rare occurrence, one last embellishment that is added to make a very predictable flip-flop is what is called *edge-triggering*. That is, when (if) a flip-flop changes state, it will do so only at the near instant of either the leadingedge or the trailingedge of the clock pulse. Both types of triggering are used (but I believe the more common choice is trailingedge). Figure 8.3.5 shows how leading-edge detection can be done by generating the ENABLE signal as the logical AND of CLK and an inverted and slightly delayed version of CLK—delayed because of the several nanoseconds of propagation time through the inverter. Beneath the circuit are timing diagrams showing a typical clock pulse, its delayed (and inverted) form, and, finally, the logical AND of those two signals. The output of the AND gate is a very brief pulse that is now the ENABLE signal in Figure 8.3.4. (This pulse will itself, of course, also experience a propagation delay through the AND gate, a delay that I've not shown in Figure 8.3.5 to keep the diagram simple.)

How does one do trailing-edge detection? With almost the same circuit as with leading-edge. If you think about what the leading-edge detector actually does, you'll see that it is detecting 0 to 1 transitions. So, if we first invert the CLK signal to convert the trailing-edge

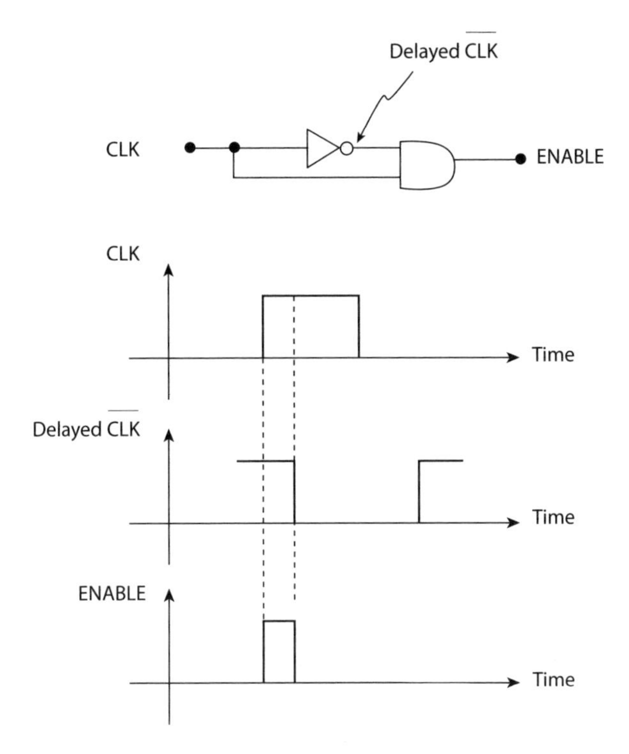

Figure 8.3.5. Leading-edge detector.

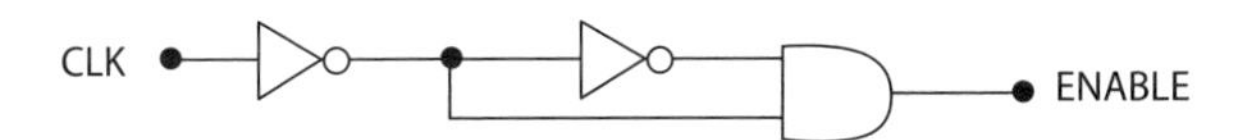

Figure 8.3.6. Trailing-edge detector.

1 to 0 transitions to 0 to 1 transitions, then we can use the circuit of Figure 8.3.5, with the result shown in Figure 8.3.6 (I'll let you sketch the diagrams to see how this works in detail). In any case, we'll now assume that edge detection, clock synchronized circuitry is part of every flip-flop and we'll use the logic symbol shown in Figure 8.3.7 to represent a clocked, edge-triggered *RS* latch (that is, an *RS* flip-flop) rather than the more detailed circuit of Figure 8.3.4. From now on, to be specific, I'll always assume *trailing*-edge detection.

The circuit of Figure 8.3.7 (compare to Figure 8.3.1) is, in fact, the solution to our original problem of getting a very brief start pulse out

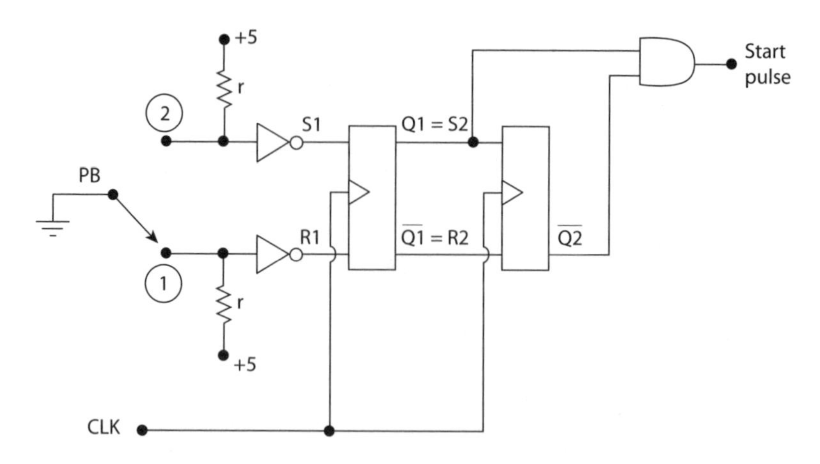

Figure 8.3.7. How to generate the start signal for a digital machine.

of a mechanical, finger-actuated push-button. The two flip-flops both automatically enter their reset states ($Q1 = 0$, $Q2 = 0$) as soon as power is turned on. The output of the AND gate, with its inputs $Q1$ and $\overline{Q2}$, is logic 0. Once *PB* is switched from contact 1 to contact 2, then at the very next clock pulse the left flip-flop sets ($Q1 = 1$), and, since $\overline{Q2}$ is still 1, the AND gate output is 1. Then, on the next clock pulse the right flip-flop is set ($Q2 = 1$) by the now set left flip-flop and the AND gate output goes to 0; that is, the AND gate output is a pulse that lasts precisely one period of the clock signal, from trailing edge to trailing edge (a one-microsecond long pulse for a 1 MHz clock). Contact bouncing has no effect, nor does how long we hold the push-button down, and the entire circuit automatically resets once the push-button is released and the switch returns to contact 1.

8.4 MORE FLIP-FLOPS

Before bringing this entire book to its technical end in the next section (the final two chapters are mostly philosophical) with a design example of a digital machine that actually does something of computational interest, let me make some final comments on flip-flops. With the addition of the *RS* flip-flop to our combinatorial logic gates, we now have all the hardware we need to build any sequential-state machine that can be built that has a finite number of states. The *RS*, however, is hardly ever used in practice, but rather other logically equivalent flip-flops are preferred. Those flip-flops are constructed by adding just a bit more embellishment to the *RS*. The three most common flip-flop variations on the *RS* are the so-called *T*, *D*, and *JK* flip-flops.

The definition of the *T* is simple. This flip-flop has the usual two outputs (Q and $\bar{Q}$), but only *one* input (T). Given the state the flip-flop is in after the trailing-edge of clock pulse n is over (that is, $Q^{(n)}$), then if $T^{(n)}=0$ at the trailingedge of clock pulse $n+1$ the flip-flop remains in that state, but if $T^{(n)}=1$ at the trailingedge of clock pulse $n+1$ the flip-flop changes to the other state. We can understand how to make a *T* from an *RS* by using the following truth table:

$T^{(n)}$	$Q^{(n)}$	$Q^{(n+1)}$	$R^{(n)}$	$S^{(n)}$
0	0	0	0,1	0
1	0	1	0	1
0	1	1	0	0,1
1	1	0	1	0

Concentrate your attention, first, on the two left-most columns ($T^{(n)}$ and $Q^{(n)}$). They show all possible combinations (four) for $T^{(n)}$ and $Q^{(n)}$. The third column shows, according to the definition of a *T* flip-flop, what $Q^{(n+1)}$ would be for each possibility. Since we are building the *T* out of an *RS*, then of course Q (and $\bar{Q}$) are the outputs of an *RS*. The final two columns of the truth table show what $R^{(n)}$ and $S^{(n)}$ must be (keeping the prohibition of $RS=1$ in mind) to give the $Q^{(n)}$-to-$Q^{(n+1)}$ state transitions that we determined in the second and third columns. Notice, in particular, that both $R^{(n)}$ and $S^{(n)}$ have a row where I've indicated there are *two* possible values, what we called "don't- care" terms back at the end of Chapter 4. Writing these don't-care terms in parentheses, as we did in Chapter 4, we can write the following Boolean equations for $R^{(n)}$ and $S^{(n)}$ in terms of $T^{(n)}$ and $Q^{(n)}$:

$$R^{(n)}=T^{(n)}Q^{(n)}+\left(\overline{T^{(n)}}\,\overline{Q^{(n)}}\right) \quad (8.4.1)$$

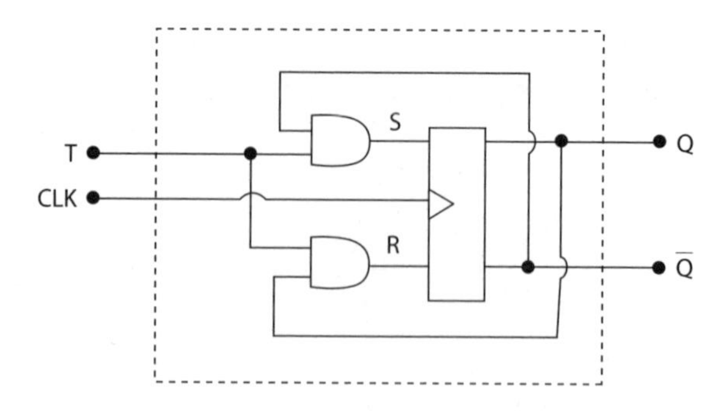

Figure 8.4.1. The T flip-flop.

and

$$S^{(n)} = T^{(n)}\overline{Q^{(n)}} + \left(\overline{T^{(n)}}Q^{(n)}\right). \tag{8.4.2}$$

A quick plotting of (8.4.1) and (8.4.2) on two-variable Karnaugh maps should convince you that the don't care terms don't result in any simplification. The final result, then, on how to build a *T* flip-flop from an *RS* flip-flop and two AND gates is shown in Figure 8.4.1.

The *D* flip-flop, like the *T*, is a one-input device. Its logical description is even simpler than is the *T*'s: whatever $D^{(n)}$ and $Q^{(n)}$ are, $Q^{(n+1)} = D^{(n)}$; that is, the *next* state is the *present* input (the *D* stands for *delay*). We can build the *D* from the *RS* just as we did the *T*. We first create a truth table showing all possible combinations of $D^{(n)}$ and $Q^{(n)}$ in the first two columns:

$D^{(n)}$	$Q^{(n)}$	$Q^{(n+1)}$	$R^{(n)}$	$S^{(n)}$
0	0	0	0,1	0
1	0	1	0	1
0	1	0	1	0
1	1	1	0	0,1

The third column for $Q^{(n+1)}$ follows from the definition of the *D* flip-flop, and from the $Q^{(n)}$ and $Q^{(n+1)}$ columns we can determine what $R^{(n)}$ and $S^{(n)}$ must be (remember, $RS \neq 1$). Writing the don't-care terms

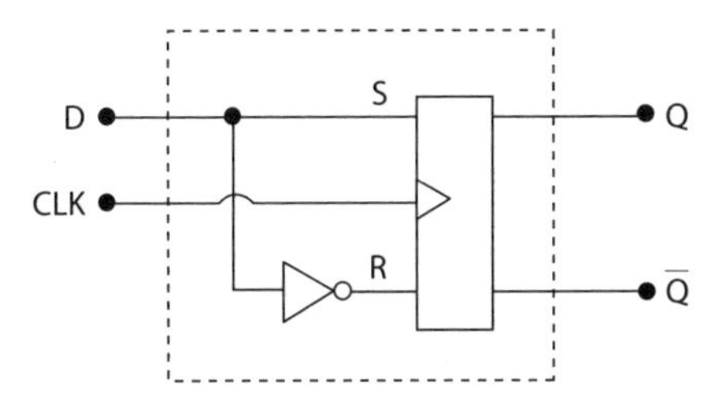

Figure 8.4.2. The D flip-flop.

in parentheses as before, we have

$$R^{(n)} = \overline{D^{(n)}}Q^{(n)} + \left(\overline{D^{(n)}}\,\overline{Q^{(n)}}\right) \tag{8.4.3}$$

and

$$S^{(n)} = D^{(n)}\overline{Q^{(n)}} + \left(D^{(n)}Q^{(n)}\right). \tag{8.4.4}$$

By inspection you should see that now, unlike in the case of the T, the don't-care terms do help, and so

$$R^{(n)} = \overline{D^{(n)}} \tag{8.4.5}$$

and

$$S^{(n)} = D^{(n)}. \tag{8.4.6}$$

To make a D from an RS, all we need to do is add one inverter, as shown in Figure 8.4.2.

And finally, the *JK* flip-flop is much like the RS (with J playing the role of S, and K the role of R—I don't believe the J and K stand for anything in particular), but without a $JK \neq 1$ restriction. If $JK = 1$ occurs, then the flip-flop changes state. Following the same process that we used for the T and D flip-flops, you should now be able to derive the circuit that converts an RS into a *JK* (try it!); the *JK* is, I believe, the most commonly used flip-flop.

8.5 A SYNCHRONOUS, SEQUENTIAL-STATE DIGITAL MACHINE DESIGN EXAMPLE

Now, to end this chapter, let's design a real machine that does something. For us, here, it will appear to be an arbitrary gadget, but you'll see it again in the next chapter where you'll discover that it is actually computing an interesting mathematical function. All that would be a distraction here, however, so for the time being just trust me that it is not an arbitrary gadget. Our machine has four states (that I'll number 0 to 3), one input (X), and two outputs (Y and Z). When we first turn power on, the machine is to start in state 1. A detailed description of what our machine is to do next, as a function of its input and state, is given by the following so-called *state-transition table*. State 0 is to be a *halting* state, and that's why there is no row for that state.

Machine State at Clock n	$X^{(n)}$	$Y^{(n)}$	$Z^{(n)}$	Machine State at Clock $n+1$
1	0	1	1	2
1	1	1	1	0
2	0	1	0	2
2	1	0	1	3
3	0	1	0	3
3	1	1	0	1

It will take 2 flip-flops to build this machine, since it takes 2 flip-flops to represent four machine states. Let's agree to use T flip-flops, which I'll call $Q1$ and $Q2$, and to let the present machine state (at clock pulse n) be represented by the binary number $Q1Q2$. For example, if $Q1$ is set ($Q1 = 1$) and $Q2$ is reset ($Q2 = 0$), the machine is in state $10 =$ machine state 2. What we need to do, then, is to find Boolean equations for $T1$ and $T2$, the inputs to the $Q1$ and $Q2$ flip-flops, respectively. From the state-transition table we can construct the following truth table, where the first five columns show how the states of the individual flip-flops change as a function of the input and the present state. By comparing $Q1^{(n)}$ and $Q2^{(n)}$ to $Q1^{(n+1)}$ and $Q2^{(n+1)}$, respectively, and knowing how a T flip-flop works (a 1 input changes the state), we can then fill in the columns for $T1^{(n)}$ and $T2^{(n)}$. The final two columns simply repeat the values of the outputs Y and Z.

$Q1^{(n)}$	$Q2^{(n)}$	X	$Q1^{(n+1)}$	$Q2^{(n+1)}$	$T1^{(n)}$	$T2^{(n)}$	Y	Z
0	1	0	1	0	1	1	1	1
0	1	1	0	0	0	1	1	1
1	0	0	1	0	0	0	1	0
1	0	1	1	1	0	1	0	1
1	1	0	1	1	0	0	1	0
1	1	1	0	1	1	0	1	0

From this table we can write the $T1$ and $T2$ equations as

$$T1^{(n)} = \overline{Q1^{(n)}}Q2^{(n)}\overline{X^{(n)}} + Q1^{(n)}Q2^{(n)}X^{(n)} \tag{8.5.1}$$

and

$$T2^{(n)} = \overline{Q1^{(n)}}Q2^{(n)}\overline{X^{(n)}} + \overline{Q1^{(n)}}Q2^{(n)}X^{(n)} + Q1^{(n)}\overline{Q2^{(n)}}X^{(n)},$$

or,

$$T2^{(n)} = \overline{Q1^{(n)}}Q2^{(n)} + Q1^{(n)}\overline{Q2^{(n)}}X^{(n)}, \tag{8.5.2}$$

Also, the output equations are

$$\overline{Y^{(n)}} = Q1^{(n)}\overline{Q2^{(n)}}X^{(n)}(\text{that is, } Y^{(n)} = \overline{Q1^{(n)}\bar{Q}2^{(n)}X^{(n)}}, \tag{8.5.3}$$

and, as the table shows by inspection,

$$Z^{(n)} = T2^{(n)}. \tag{8.5.4}$$

That's it![3] You are now (theoretically) able to design *any* synchronous digital machine that *can* be built.

NOTES AND REFERENCES

1. In the construction of a digital machine, the end-product from the design engineer is what is called the *wire-list*. That is a detailed description of what is connected to what, to be used by the technicians on the production floor who are the people that actually put the machine together. To prepare a wire-list it is necessary to give every wire a name, and in particular the wire out of the start-button circuit (what is called the "start pulse" in Figure 8.3.7) will have a name. In a 1963 machine, the first I designed, I named that wire GBG, for *Go Baby Go*—not particularly imaginative, but certainly descriptive of what I prayerfully remember hoping would happen when I first pushed the start button! (For the historical record, the first time I pushed that button the machine did go, for a while. Then it stumbled all over its "feet" and came to a temporarily discouraging halt. There then followed a fairly lengthy debugging phase until, at last, the machine ran flawlessly. Or anyway, that's my story of how those long-ago events evolved, and I'm not budging from it!)

2. When I was a student at Stanford in EE 266 (see note 1 in Chapter 2), I don't recall hearing any mention of bouncing switches. All the switches that appeared on homework and exam problems were perfect switches. I think realistic, bouncing switches are, today, routinely discussed even in introductory logic circuit courses.

3. Well, not quite. We are not completely done until we make sure that our machine starts in state 1. Before the machine begins to look at the input X, we need to reset $Q1$ ($Q1 = 0$) and to set $Q2$ ($Q2 = 1$), which could be done by applying the output signal from Figure 8.3.7 directly to $R1$ and $S2$, respectively, of the RS flip-flops from which we built the T flip-flops used in our machine.

9

Turing Machines

No, I'm not interested in developing a *powerful* Brain.
All I'm after is just a *mediocre* brain, something like the
President of the American Telephone and
Telegraph Company.

Shannon wants to feed not just *data* to a Brain, but *cultural*
things! He wants to play music to it!
—Both by Alan Turing, during a two-month visit in early 1943 to
Bell Labs (New York City), where he met Claude Shannon and
found they had a common interest in how computing machines
might imitate the human brain

A very small percentage of the population produces the
greatest proportion of the important ideas. This is akin to an
idea presented by the English mathematician, Turing,
that the human brain is something like a piece of
uranium . . . You shoot one neutron into it [and more than
one neutron is] produced [the famous *chain-reaction*].
Turing says this is something like ideas in the human brain.[1]
There are some people if you shoot one idea into the brain,
you will get [alas!] half an idea out. There are other people
who . . . produce two ideas for each idea sent in.
—Claude Shannon, in a March 1952 talk at Bell Labs on
creativity, during which he explained how he arrived
at the logic circuitry for a machine that plays a perfect
game of Nim

9.1 THE FIRST MODERN COMPUTER

A *Turing machine* is the combination of a sequential, finite-state machine plus an external read/write memory storage medium called the *tape* (think of a ribbon of magnetic tape). The tape is a linear sequence of squares, with each square holding one of several possible symbols. Most generally, a Turing machine can have any number of different symbols it can recognize, but I'll assume here that we are discussing the 2-symbol case (0 or 1). In 1956, Shannon showed that this in no way limits the power of what a Turing machine can do.

The tape is arbitrarily long in at least one, perhaps both, directions. The finite-state machine is connected to a *read/write head*, which at each *machine cycle* (I'll define what that is in just a moment) is located over a square on the tape. The head does three distinct operations during a cycle (these three operations, in fact, together with a final, fourth operation, *define* a machine cycle): first, the head reads the symbol on the square it is over, then it overwrites that symbol (perhaps with the same symbol), and then the head moves at most one square (that is, to either the left or to the right neighbor square, or it doesn't move and so remains over the current square). Depending both on its present state and the symbol just read, the fourth and final operation of a machine cycle occurs when the finite-state machine transitions to a new state (which may, in fact, be the present state). Then a new machine cycle begins. Figure 9.1.1 shows the connection of a finite-state machine, the read/write head, and the tape. The entire arrangement, all together, is what we call a Turing machine.

When placed into operation we'll imagine that the tape is initially blank (that is, the symbol 0 is on all of the tape's squares)—except for some finite number of squares that have 1s. By convention, we'll always take the finite-state machine as initially in state 1. And finally, we must specify over which square on the tape the read/write head is initially placed. The finite-state machine and the read/write head then move along the tape (we imagine the tape is motionless) according to the internal details of the finite-state machine and the particular sequence of symbols encountered on the tape. At some time after we turn the Turing machine on, it presumably completes its task (whatever that might be), and the finite-state machine enters state 0 (called the *halting state*) and stops.

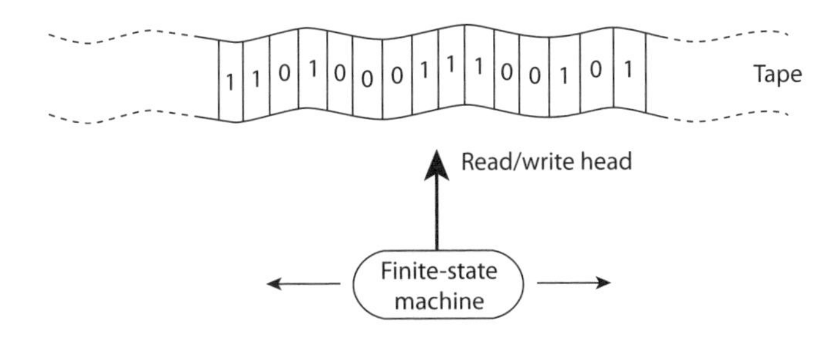

Figure 9.1.1. A Turing machine.

As far as I know, nobody has ever actually constructed a Turing machine. It is a purely theoretical concept. Described in 1936 by the English mathematician Alan Turing (see Shannon's mini-biography in Chapter 3)— that is, ten years before the first actual electronic computers began to be constructed—Turing's machines are nevertheless as powerful as any modern machine in what they can compute. Compared to a modern, electronic-speed computer, a Turing machine is really, really slow (think of a race between a snail and a photon), but time is irrelevant. There is an infinity of it yet to come. Given enough time, a Turing machine can compute anything that *can* be computed.[2]

A Turing machine's power to compute comes not from super technology, but from its tape, for two reasons. First, Turing was the first to conceive of the idea of a *stored program* that could be changed *by the operation of the machine itself*. The program, and its input data, exist together on the tape as sequences of symbols. And second, because of the arbitrarily long length of the tape, a Turing machine has the ability to "remember" what has happened in the arbitrarily distant past.

In developing this view of a computing machine, Turing was not suggesting it as a practical design for an actual machine. Rather, as a mathematician he used his machines as a conceptual framework in which to study the limits on just what mechanistic devices *can* actually compute. Indeed, the title of his 1936 paper, "On Computable Numbers, with an Application to the Entscheidungsproblem"—that final tongue-twister translates as "the decision problem"—clearly shows Turing's intent. His great accomplishment was to show that not all the numbers we can imagine are in fact actually computable. That is, Turing showed there are limits to what a computer—any computer—can do. I'll return to the concept of a computable number in the final section of this chapter.

9.2 TWO TURING MACHINES

As my first specific example of a Turing machine, consider the following state-transition table for a Turing machine that adds any two non-negative integers m and n that are placed on the tape, leaves their sum on the tape, and then halts. Our convention for representing m and n is to write $m+1$ consecutive 1s for m, then a 0, and then $n+1$ consecutive 1s for n. All the rest of the tape squares, initially, have the symbol 0. (This representation for m and n is called *unary* notation.) The reason for using one additional 1 for each value of m and n is to allow either (or both) of m and n to be zero (which is represented by a single 1). This Turing machine is said to *compute the function* $f(m,n)=m+n$. Notice that in this machine the read/write head either doesn't move during a machine cycle or, if it does move, it does so always to the right. Since this Turing machine has six states (including the halting state), it would take three flip-flops to build the finite-state machine portion. The first row of the table can be read as: if in state 1 and reading a 0, then write a 0, move right (R), and enter state 1. The second row can be read as: if in state 1 and reading a 1, then write a 0, don't move (−), and enter state 2. And so on.

Present State	Tape Symbol	Operation	Next State
1	0	0/R	1
1	1	0/-	2
2	0	0/R	2
2	1	0/-	3
3	0	0/R	3
3	1	1/R	4
4	0	1/-	5
4	1	1/R	4
5	0	halt	0
5	1	1/R	5

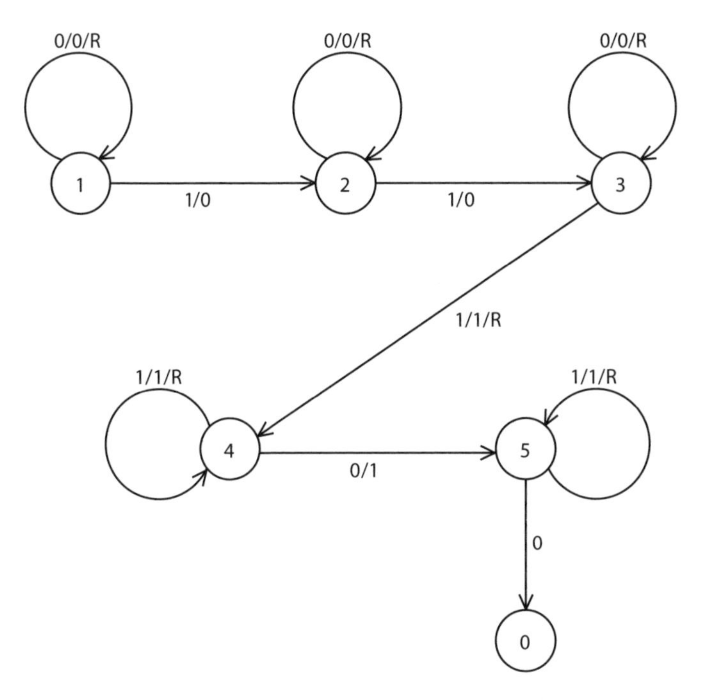

Figure 9.2.1. The state-transition diagram for a Turing machine adder.

It may not be at all obvious from the state-transition table how the machine works, but using the table to draw the state-transition diagram of Figure 9.2.1 should make the operation a lot more transparent. The circles with numbers inside are the states, and the curved lines with arrowheads represent the transitions from state to state, with each directed line marked with the conditions causing that transition. For example, 0/0/R means "read 0, write 0, move right." State 0 is a halting state. This machine is so simple that, starting in state 1 with the read/write head over any tape square to the left of the first 1, you should be able to follow the state-transition diagram step by step, with pen and paper, to confirm the following three test cases:

$$0+2=\cdots 0101110\cdots \text{which should give}\cdots 01110\cdots$$

$$2+0=\cdots 0111010\cdots \text{which should give}\cdots 01110\cdots$$

$$2+3=\cdots 0111011110\cdots \text{which should give}\cdots 01111110\cdots .$$

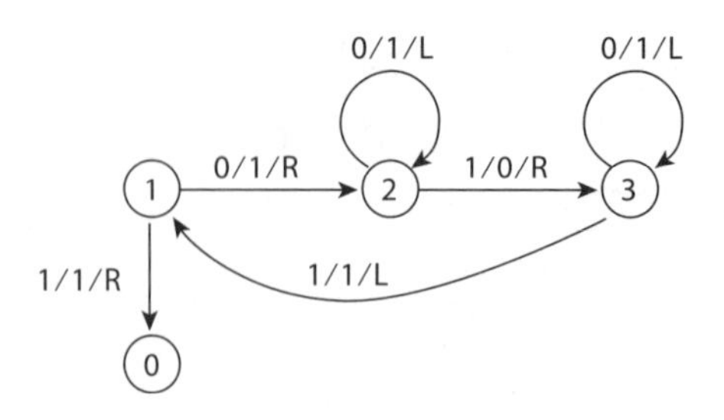

Figure 9.2.2. Radó's Busy Beaver Turing machine.

Computer scientists can be as playful as anybody, and for my second example (which I think supports that claim) consider the Turing machine with the state-transition diagram of Figure 9.2.2. (The "L" means, of course, "move left.") This machine was "invented" by the Hungarian-born American mathematician Tibor Radó (1895–1965) in 1962 to illustrate what he called the "Busy Beaver Game." For a given number of states k (not counting the halting state), the "game" is to find the state-transition diagram that results in the production of a tape with the most 1s on it when the machine halts, starting with an initial tape of all 0s. Another condition set by Radó is that the read/write head *must* move on each machine cycle, that is, the option of remaining over the tape square just scanned is not available to a potential Busy Beaver. There is no rest for a Busy Beaver!

If you look closely at Figure 9.2.2, and compare it with the state-transition table of the machine we designed with T flip-flops at the end of the last chapter, you should notice that they are one and the same. In that machine, the input X is the symbol read, the output Y is the symbol written, and the output Z is the read/write head's motion control ($Z = 1$ is R and $Z = 0$ is L).

Radó called a "maximum-production of 1s" machine the *k-state, 2-symbol Busy Beaver*. Writing $\sum(k)$ to denote the number of 1s on its tape when the k-state, 2-symbol Busy Beaver halts, it is known today that $\sum(3) = 6$ (when he wrote in 1962 Radó didn't know the value of $\sum(3)$—and his machine in Figure 9.2.2 doesn't achieve it—and he felt that the discovery of the value of $\sum(4)$ to be "entirely hopeless at present"). With the challenge thrown down, however, it didn't take long (1966) for it to be shown that $\sum(4) = 13$. Radó also defined the function $S(k)$ as the number of moves (or steps) made by the read/write head of a k-state, 2-symbol Busy Beaver Turing machine from start to finish. Can you figure out how many moves Radó's machine makes, as well as how many 1s it writes on its tape before it halts? If not, or to check your answers, see the notes.[3]

The values of $\sum(k)$ and $S(k)$ for all k ≥ 5 Busy Beavers are still unknown, although lower bounds have been determined. For example, as I write (in 2011) it is known that $\sum(5) \geq 4{,}098$ and $S(5) \geq 47{,}176{,}870$, and that $\sum(6) \geq 10^{10{,}566}$ and $S(6) \geq 10^{21{,}132}$. And, to really give you pause for stupified wonder,

$$\sum(10) \geq 3^{3^{3^{\cdot^{\cdot^{\cdot^{3}}}}}}$$

where there are $3^{27}(= 7{,}625{,}597{,}484{,}987)$ 3s in the exponential stack! As you can see, Radó's $\sum(k)$ and $S(k)$ are functions that grow at stupendously fantastic rates as k increases. It is known, in fact, that both $\sum(k)$ and $S(k)$ are what computer scientists and mathematicians call *noncomputable functions*. The concept of computability is a very deep one, and I'll simply brush the surface of it in the next section.

But, before we get to that, a couple of final comments on Turing machines. First, Shannon suggested that a reasonable measure of the "complexity" of a Turing machine is the product of the number of different symbols that can appear on the tape, and the number of states for the finite-state machine portion. That is, symbols and states can be traded-off against each other. In 1956, in fact, Shannon took this trade-off to the limit and proved the remarkable result that if one uses enough symbols, then, given any computable function, there exists a Turing machine with just two states that can evaluate that function. A companion demonstration shows that if one uses enough states, then, for any computable function, there exists a Turing machine using just two symbols.

And second, one of the most astonishing results in Turing's 1936 paper is that we do not have to build a different Turing machine for every new function we wish to compute. He showed that there exists a so-called *Universal Turing Machine* (UTM) that accepts as its input (that is, what's initially written on its tape) a description of any other specific Turing machine (that description includes the state-transition table and the initial tape of the specific Turing machine) and then the UTM simulates the specific machine. That is, the details of each new function can be completely absorbed by the UTM tape, with the finite state portion of the UTM unchanging. You might suspect that a UTM would be very complicated, but that's not so. It can be realized, for example, as a 6-symbol, 21-state (plus a halting state) machine.[4]

9.3 NUMBERS WE CAN'T COMPUTE

This book has been pretty "engineery" up through the first eight chapters, with this chapter being the first to get "philosophical." What else, after all, would you call any discussion about a fantastic gadget like a Turing machine with its infinite memory! Well, in this final section of a philosophical chapter we'll really go extreme in the philosophizing department. You may, at first, wonder where I'm going with it, but just stick with me and I think you'll find it worth your time. What I'm going to show you is that, given any computer of finite size, even if it is of absolutely gargantuan size (say 10^{100} flip-flops and a memory of $10^{10^{100}}$ bits, that is, a googol of flip-flops and a googelplex of bits), there is an uncountable infinity of numbers our computer can *not* compute. And that will be true even if you run the computer at a clock frequency of a thousand, million, trillion gigahertz (a value that will never be achieved, and in the final chapter I'll show you why not) for a thousand, million, billion, trillion centuries.

First, a few observations about the concept of infinity. We all know it's "big," but that doesn't even begin to get at the mathematics of what it means to say something is *infinite* in size. Most people, when asked to give an example of an infinite number of things will reply "All the integers." That's correct, too, as the integers form what mathematicians call a *countably infinite set*. We can literally count the integers, one, two, three, . . . , up to a billion, a trillion, and on and on. If we count off one integer each second, then I can tell you precisely when in the future I'll have counted up to any specific integer you name. The integers are infinite in number, yes, but they can be *counted*. This is probably not so surprising to you, but there are other countable infinities, and they *are* surprising. For example, all the rational numbers (that is, all the numbers that are the ratios of integers) from zero to infinity are a countable infinity. Why is that a surprising statement?

It is surprising because, unlike the integers, the rationals are *dense*. What a mathematician means by that is that if you take any two fractions, no matter how close together they may be, there is an infinity of rationals between them. And between any two of that infinity of rationals there is yet another infinity of rationals. And so on, forever. The integers are *not* dense because there is a minimum separation between them of—of course!—one. There is no minimum separation between the rationals.[5] Nevertheless, despite their denseness, the rationals are still a countable infinity. In other words, the rationals and the integers are *infinite sets of the same size*, even though the set of the integers is included in the set of rationals. This astonishing result, totally at odds with intuition (what is often called "common sense"), was discovered by the Russian-born German mathematician Georg Cantor (1845–1918) in 1874. Most mathematicians of his day thought Cantor was crazy, but it was they (not him) who were wrong (although, ironically, Cantor died in a mental institution).

To show how the rationals are countably infinite, Cantor had two brilliant ideas. First, he showed how to systematically write down all the rationals so as not to overlook even one. He did that in the form of an infinitely large two-dimensional matrix array. Second, he showed how to count through that infinite matrix in such a way as to be able to tell you precisely when he would reach any specific rational number you might name. Figure 9.3.1 illustrates both of Cantor's ideas.

The first (infinitely long) row in Figure 9.3.1 gives all the rationals with a denominator of 1, that is, all the integers. The second (infinitely long) row gives all the rationals with a denominator of 2, the third (infinitely long) row gives all the rationals with a denominator of 3, and so on. There is a lot of repetition in the matrix (indeed, *every entry* along the main diagonal is 1), but that's okay — the point is that

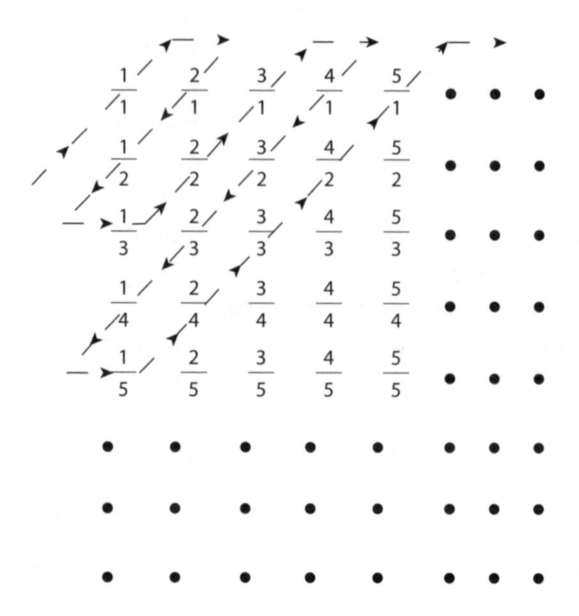

Figure 9.3.1. Cantor's infinite matrix of the rationals.

the matrix includes *every* rational number between zero and infinity. This was Cantor's brilliant first idea. If we now make the mistake of counting-off the rationals along the rows, however, we'll never get off of the first row! The same problem occurs if we try to count down columns. Cantor's second brilliant idea was to count along *diagonal* paths that weave back-and-forth through the matrix, as shown by the dashed line with the arrow heads in Figure 9.3.1. *Then* we *will* arrive, in finite time, at any rational in the matrix that you care to specify. For example, the first rational is $\frac{1}{1}$, the third rational is $\frac{1}{2}$, the ninth rational is $\frac{2}{3}$, and so on. The rationals are, therefore, a countable infinity.

You might now be wondering if *all* infinite sets are countable. The answer is *no*, there are indeed infinite sets that can't be counted. The easiest to understand is the set of all real numbers and, again, it was Cantor with a brilliant third idea who showed this. Once we have this result in hand, we'll have all we need to show the existence of noncomputable numbers. Cantor accomplished his proof by the classic approach of assuming the opposite of what he wanted to show and then deriving a contradiction. That is, Cantor's proof begins by assuming the real numbers from zero to one *are* a countable infinity, which means we can list them, one after the other, in some order. So, suppose that list is, in decimal notation,

$$\begin{bmatrix} d_1 = & 0. & d_{11} & d_{12} & d_{13} & d_{14} & \cdots \\ d_2 = & 0. & d_{21} & d_{22} & d_{23} & d_{24} & \cdots \\ d_3 = & 0. & d_{31} & d_{32} & d_{33} & d_{34} & \cdots \\ d_4 = & 0. & d_{41} & d_{42} & d_{43} & d_{44} & \cdots \\ . & . & & & & & \cdots \\ . & . & & & & & \cdots \\ . & . & & & & & \cdots \end{bmatrix}$$

where d_{ij} is the jth decimal digit of d_i, the ith real number. Cantor's brilliant third idea was how to show that there exists a number, $n = 0.n_1n_2n_3n_4\ldots$, where n_i is a decimal digit of n, which is not on this supposed list, and so the list is not complete, and so we have in fact failed to count all the real numbers.

There are various versions of Cantor's idea (which is, again, a diagonal argument), but here's the essential nature of it. Starting with the first number on the supposed list, d_1, ask if $d_{11} = 5$? If so, write $n_1 = 4$, and if not write $n_1 = 5$. Then, moving to the next number on the supposed list, d_2, ask if $d_{22} = 5$? If so, write $n_2 = 4$, and if not write $n_2 = 5$. Then, moving to the next number on the supposed list, d_3, ask if $d_{33} = 5$? If so, write $n_3 = 4$, and if not write $n_3 = 5$. And so on. The number n that we get from all this, $0.n_1n_2n_3n_4\cdots$, can *not* be in the list because n $\neq d_1$ since their first decimal digits are different (by construction), and n $\neq d_2$ since their second decimal digits are different (by construction), and n $\neq d_3$ since their third decimal digits are different (by construction), and so on. End of proof! The real numbers are the elements of an uncountable infinite set, an infinity that Cantor called the *continuum*.

In a 1905 summer lecture at the University of Göttingen, the great German mathematician David Hilbert (1862–1943) proclaimed Cantor's two results to be the basis for "one of the most beautiful proofs in set theory." And after Cantor's death Hilbert wrote (in a letter to a colleague) that "for originality and boldness of thought, there is no mathematician in history—from Euclid to Einstein—who surpassed him."

Now, if we combine Cantor's two results, the countable infinity of the rationals and the uncountable infinity of the reals, the inexorable conclusion we are forced to make is that there is an uncountable infinity of real numbers that are *not* rational. We (of course) call that uncountable infinite set the *irrationals* (numbers that include, for example, $\sqrt{2}$ and π). It is amusing to note that while the irrationals are "nearly all" of the reals, it is generally not at all easy to prove that any particular irrational actually *is* irrational! And now, at last, we are ready to prove the result I promised you, the existence of noncomputable numbers.

We start by imagining our huge computer being programmed to compute numbers using various algorithms, the details of which are unimportant. The programming is done in any language you wish (to understand this argument you don't even have to know a programming language!), with the only requirement being that the language uses a finite set of symbols. For example, the twenty-six letters of the English alphabet, the ten digits from 0 to 9, and a few extra special symbols like >, <, =, ^, (,), and so on. We suppose each new program we write computes a new number.

Let's now list all of these programs by length. That is, the first program on our list will have one symbol to it. If there is more than one program of length one, we'll list them in alphabetical order. (The number of programs of a given, fixed length is, of course, finite, since we have a finite set of symbols at our disposal.) Then come all the programs of length 2 (sorted alphabetically), then all those programs of length 3 (sorted alphabetically), and so on. Clearly, all possible programs will form a *countable* infinity. But there is also an *un*countable infinity of numbers, and so there must be an uncountable infinity of numbers left uncomputed, numbers we simply can't compute because there simply aren't enough programs. Notice, carefully, that we have not identified any particular number that can't be computed, just that there are an infinity of them. Further, this argument does not necessarily associate the countable infinity of computable numbers with the countable infinity of the rationals. All we have established is that each set is a countable infinity, not that they are the same set.

One possible objection to this argument might be to the limitation of one computed number to a program. What if we instead imagine that each of our countably infinite programs could somehow calculate more than one number, perhaps hundreds or even millions of numbers. Would that make a difference? No. In fact, if each program could somehow calculate a countable infinity of numbers, the total number of computed numbers would still be "just" a countable infinity. That's because

(countable infinity) times (countable infinity) = countable infinity.

For example, what do you get if you multiply all the rational fractions by all the rational fractions? Why, nothing more or less than just all the rational fractions back again![6]

NOTES AND REFERENCES

1. The reference to Turing is almost certainly due to Shannon having read Turing's famous paper "Computing Machinery and Intelligence," *Mind*, October 1950, pp. 433–460. It was in this paper that Turing put forth what was to become famous in computer science as the *Turing test*, an experimental procedure to unemotionally decide if a machine possessed artificial intelligence. For Turing's comparison of ideas to neutrons, see in particular, p. 454.

2. MIT electrical engineering professor Marvin Minsky refers to this issue in his beautiful book *Computation: Finite and Infinite Machines*, Prentice Hall, 1967, p. 128. Despite what he writes as the "staggering inefficiency" of a Turing machine, Minsky goes on to say, "It is possible to execute the most elaborate possible computational procedures with Turing machines whose fixed structures [that is, the finite-state machine and the read/write head] contain only dozens of parts [this excludes the arbitrarily long tape if we count each writable/erasable square as a distinct 'part']. One can imagine an interstellar robot, for whom reliability is the prime consideration, performing its computations in such a leisurely manner, over eons of spare time." Historical note: It was Minsky, while working at Bell Labs in the summer of 1952, who gave Shannon the idea for the "Ultimate Machine" that so fascinated Arthur C. Clarke in Chapter 3.

3. Radó's machine puts 5 1s on its tape in 21 moves before halting. You can confirm this with the simple-minded (but also *very* tedious) method of following along with pen and paper as the state-transition diagram of

Figure 9.2.2 bounces back and forth among its three states. Far easier on your brain, however, is to write a computer simulation program to generate what the tape looks like at the end of each machine cycle time. That's what I did, with the following code called **rado.m**. It's in MATLAB, but the program is so elementary I think it would be trivial to rewrite it in any of the common programming languages; *for/end* and *if/elseif/end* loops work the same, independent of language. The only MATLAB functions that may need some explanation are *zeros* (which sets each of the 50 elements of the row vector *tape* equal to zero, initially), and *sum* (which adds the 50 elements of *tape*, giving the number of elements of *tape* equal to 1). Both operations can easily be implemented as *for/end* loops in other languages. The very last line of **rado.m** prints the final values of *shift* and the number of 1's in *tape*.

---

rado.m

```
state=1;tape=zeros(1,50);location=25;shift=0;
while state>0
    symbol=tape(location);
    if state==1
        if symbol==0
            tape(location)=1;location=location+1;state=2;
        else
            tape(location)=1;location=location+1;state=0;
        end
    elseif state==2
        if symbol==0
            tape(location)=1;location=location-1;state=2;
        else
            tape(location)=0;location=location+1;state=3;
        end
    else
        if symbol==0
            tape(location)=1;location=location-1;state=3;
        else
            tape(location)=1;location=location-1;state=1;
        end
    end
    shift=shift+1;
end
sum(tape),shift
```

---

4. See Minsky's book (note 1), Chapter 7, pp. 132–145, and in particular Figure 7.2-9 on p. 142, for the state-transition diagram of a UTM. The symbols

are 0, 1, X, Y, A, and B (the figure shows two additional symbols, M and S, but in a later note Minsky says each can be replaced with an X).

5. For example, suppose we start with the two rational fractions $\frac{1}{6}$ and $\frac{1}{7}$. The fraction halfway between them is

$$\frac{1}{7}+\frac{1}{2}\left(\frac{1}{6}-\frac{1}{7}\right)=\frac{1}{7}+\frac{1}{84}=\frac{13}{84},$$

another rational fraction. We can continue this bisecting process as long as we wish, finding ever more rational fractions as close together as we wish.

6. What I've discussed here concerning infinity is just the beginning. Cantor also showed that there are "as many" points (numbers) in the interval 0 to 1 as in the entire real line from minus infinity to plus infinity. Indeed, the continuum is the same uncountable infinity as the number of points in a space of any finite or countably infinite dimension: in particular, the points in the interiors of the unit square or the unit cube can be put into a one-to-one correspondence with the points in the unit interval. Even Cantor himself was astonished at that discovery. In a June 1877 letter to the German mathematician Richard Dedekind (1831–1916) he wrote of it, "I see it, but I don't believe it!" (Cantor and Dedekind had an extensive correspondence on infinity, and you can find more on their interaction in Fernando Q. Gouvêa, "Was Cantor Surprised?" *American Mathematical Monthly*, March 2011, pp. 198-209.) There is an infinite number of ever larger infinities *beyond* the continuum, too, the so-called *power sets of Cantor* (the set of all subsets of an infinite set). That is all beyond the scope of this book, but if you're interested in reading more I can recommend Theodore G. Faticoni, *The Mathematics of Infinity*, Wiley-Interscience, 2006; and Eli Maor, *To Infinity and Beyond*, Birkhäuser, 1987 (reprinted in 1991 by Princeton University Press).

10

Beyond Boole and Shannon

0101100101100101100010

Since the output is implicit in the input, no computation ever generates information.
—Charles Bennett and Rolf Landauer, *pioneers in the realization that classical Boolean logic gates actually destroy information*

10.1 COMPUTATION AND FUNDAMENTAL PHYSICS

In a previous book, I opened the first section ("The Limits of Computation") of the final chapter with these words:

> The speed of any computer is fundamentally limited by how fast its various component parts can send and receive signals among themselves—that is, by the speed of light and by how far those signals have to travel. We can't do anything about the speed of light, but one way to increase the speed of a computer is simply to make it smaller. That means the computer's volume decreases. Suppose, just to be specific, a computer has the shape of a sphere with radius r. As $r \to 0$, the volume decreases as r^3. If the energy required to power our shrinking computer doesn't decrease at least as fast, then the dissipated heat energy density in the computer will increase and the temperature of the computer will rise. Eventually, the computer will melt. That's one sort of limitation on computers, of a *physical* nature, one that can be gotten around—at least for a while—by various means (the most obvious being to add a cooling system to the computer). There are other limitations, however, of a far more profound nature.[1]

In that book I discussed one of those "other limitations," of a *mathematical* nature, the famous "halting problem."[2] (The issue of computability that we discussed in the last chapter is also one of pure mathematics.) I did not, however, discuss in that previous book any physical limitations on computers, other than the one above; so, in this final chapter of this book I want to briefly touch on how fundamental physics—the uncertainty principle from quantum mechanics, and thermodynamics, for example—constrain what is possible, in principle, for the computers of the far future. The general conclusion will be that while there are indeed finite limitations, present-day technology falls so far short of those limits that there will be good employment for computer technologists for a very long time to come.

Certainly Boole, and as far as I know Shannon, too, never studied the physics of computation. Obviously Boole simply couldn't have, as none of the required physics was even known in his day, and Shannon was nearing the end of his career when such considerations were just beginning. And yet, you'll see that we will use both Boole's algebra and Shannon's information concepts to make many of our calculations. Both men would, I think, have found what we'll do next fascinating.

The very concepts of quantum mechanics would have seemed to be mumbo-jumbo to Boole, but thermodynamics was well on its way to becoming a mathematical science when Boole was still a child. In 1824 the French engineer Sadi Carnot (1796–1832) published his groundbreaking book *Reflections on the Motive Power of Fire*, in which he studied how a heat engine (for example, a steam engine) could achieve maximum efficiency. If he had been interested in such matters (I suspect not, but who really knows?), Boole could surely have mastered Carnot's book. But the *last* thing he would have thought to do was to apply Carnot's engineering arguments to mathematical logic. As one writer put it so nicely. "In the nineteenth century, despite the vision of Babbage,[3] computation was thought of as a mental process, not a mechanical one. Accordingly, the thermodynamics of computation, if anyone had stopped to wonder about it, would probably have seemed no more urgent as a topic of scientific inquiry than, say, the thermodynamics of love."[4]

10.2 ENERGY AND INFORMATION

In a classic paper by two pioneers in the physics of computation, the American physicists Charles Bennett (born 1943) and the German-born Rolf Landauer (1927–1999), we read "Information is destroyed whenever two previously distinct situations become indistinguishable."[5] Their first example of what is meant by that comes from an everyday observation we've all made at some time. Two rubber balls dropped from different heights are, by definition, in two initially distinct states, but after a time they both end up in the same state; after perhaps numerous bounces, each will come to rest on the ground. The reason for that is, of course, energy is dissipated at each bounce. Observation of the final state does not allow us to determine the initial states, and so we conclude that the energy dissipation has resulted in a loss of information. As a second example of a loss of information, this time in a situation obviously computational, Bennett and Landauer ask us to imagine that we are told that the result of adding two non-negative integers is 8. That's information, yes, but not as much as knowing what those two integers were (was it $6 + 2$, or $4 + 4$, or $5 + 3$, or ...?) That is, there was a certain amount of information going *into* the adder, but less information coming *out*. This is called an *irreversible computation*, in that we cannot reconstruct the inputs from knowledge of the output.

A similar loss of information occurs in digital circuits constructed with Boolean logic gates. To be specific, consider the truth table for a two-input (A and B) NAND gate, with the output F:

A	B	F
0	0	1
0	1	1
1	0	1
1	1	0

Let's suppose the two logical inputs are each equally likely to be 0 or 1. I probably could safely assume that you'd believe me, without proof, that this represents an information input to the NAND gate of 2 bits, but let's calculate it anyway using Shannon's formula (7.1.2).

For the inputs, we have four possibilities (00, 01,10, and 11), each with probability $\frac{1}{4}$. Thus, the *input information entropy*, H_{in}, is given by

$$H_{in} = 4\left[-\frac{1}{4}\log_2\left(\frac{1}{4}\right)\right] = -\log_2\left(\frac{1}{4}\right) = \log_2(4) = \log_2(2^2) = 2\text{bits},$$

just as expected. There are just two possibilities for the output F, however (0 and 1), and they are not equally-likely. That's because the 0 output occurs only for one of the four possible inputs (and so $F = 0$ has probability $\frac{1}{4}$), while the 1 output occurs for each of the other three possible inputs (and so $F = 1$ has probability $\frac{3}{4}$). Thus, the *output information entropy*, H_{out}, is given by

$$\begin{aligned} H_{\text{out}} &= -\frac{3}{4}\log_2\left(\frac{3}{4}\right) - \frac{1}{4}\log_2\left(\frac{1}{4}\right) = \frac{3}{4}\log_2\left(\frac{4}{3}\right) + \frac{1}{4}\log_2(4) \\ &= \frac{3}{4}\left[\log_2(4) - \log_2(3)\right] + \frac{1}{4}\log_2(4) = \log_2(4) - \frac{3}{4}\log_2(3) \\ &= 2 - \frac{3}{4}\log_2(3)\,\text{bits} = H_{\text{in}} - \frac{3}{4}\log_2(3)\,\text{bits}. \end{aligned}$$

We thus see that $H_{\text{out}} < H_{\text{in}}$ by the amount of

$$\frac{3}{4}\log_2(3)\,\text{bits} = 1.189\,\text{bits}.$$

The NAND gate has destroyed more than half of the input information! Similar calculations for the AND, OR, and XOR gates will show information losses, as well. Only the NOT gate will preserve information, and you'll notice that with the NOT function we can reconstruct the input from the output. The NAND, OR, AND, and XOR gates are *logically irreversible* gates, while the NOT gate is a *logically reversible gate*.

10.3 LOGICALLY REVERSIBLE GATES

To build logically reversible gates that allow any Boolean function to be built, not just the NOT, is not at all difficult. A key observation is that there must be as many output lines as there are input lines (a condition satisfied by the NOT). A little reflection should make this obvious; if there are more inputs than outputs, then the same output must occur for more than one combination of inputs—and that's the cause of information loss! (If there are fewer inputs than outputs, then the same input combination would have to cause more than one output, and now we talking about an *unpredictable* gate!) One such logically reversible gate is the *Toffoli gate*, named after its inventor, Tommaso Toffoli (born in Italy in 1943 and now professor of electrical and computer engineering at Boston University). Put forth in 1980, it has three inputs (A, B, C) and three outputs (A, B, C′) with the following truth table:

A	B	C	A	B	C'
0	0	0	0	0	0
0	0	1	0	0	1
0	1	0	0	1	0
0	1	1	0	1	1
1	0	0	1	0	0
1	0	1	1	0	1
1	1	0	1	1	1
1	1	1	1	1	0

The two input variables A and B are called *control* inputs because their values determine the value of the output variable C'. (The values of the two control variables A and B never change from input to output.) In a Toffoli gate, $C' = \bar{C}$ if $A = B = 1$. Otherwise, $C' = C$. For this reason a Toffoli gate is often called a *controlled-controlled-NOT (or CCN) gate*. (Recall the controlled-NOT, or CN gate, otherwise known as the XOR, which was used in the Hamming code error correction logic of Figure 7.5.4.) Notice, in particular, that if $C = 1$ then $C' = \overline{AB}$

and so this special case of the Toffoli gate is a NAND gate. Since we know we can build any Boolean function from NANDs, we also know we can build any Boolean function from Toffoli gates. From the truth table for the Toffoli gate we can write

$$C' = \bar{A}\bar{B}C + \bar{A}BC + A\bar{B}C + ABC̄,$$

which, when plotted on a three-variable Karnaugh map, immediately shows us that

$$\begin{aligned} C' &= AB\bar{C} + \bar{B}C + \bar{A}C = AB\bar{C} + (\bar{A} + \bar{B})C \\ &= AB\bar{C} + \overline{AB}C = AB \oplus C, \end{aligned}$$

which means we can logically realize a Toffoli gate from an AND and an XOR.

There is a mathematically nice way, using matrix notation, to express the operation of a Toffoli gate. (I'm doing this with the understanding that simple matrix ideas are now taught in high school.) We can write the gate's truth table as follows, with the 8 by 3 output matrix on the right in (10.3.1) as the result of an 8 by 8 matrix **T** "operating on" the 8 by 3 matrix on the left in (10.3.1). That is, if

$$T = \begin{bmatrix} 1 & 0 & 0 & 0 & 0 & 0 & 0 & 0 \\ 0 & 1 & 0 & 0 & 0 & 0 & 0 & 0 \\ 0 & 0 & 1 & 0 & 0 & 0 & 0 & 0 \\ 0 & 0 & 0 & 1 & 0 & 0 & 0 & 0 \\ 0 & 0 & 0 & 0 & 1 & 0 & 0 & 0 \\ 0 & 0 & 0 & 0 & 0 & 1 & 0 & 0 \\ 0 & 0 & 0 & 0 & 0 & 0 & 0 & 1 \\ 0 & 0 & 0 & 0 & 0 & 0 & 1 & 0 \end{bmatrix}$$

then

$$\mathbf{T}\begin{bmatrix} 0 & 0 & 0 \\ 0 & 0 & 1 \\ 0 & 1 & 0 \\ 0 & 1 & 1 \\ 1 & 0 & 0 \\ 1 & 0 & 1 \\ 1 & 1 & 0 \\ 1 & 1 & 1 \end{bmatrix} = \begin{bmatrix} 0 & 0 & 0 \\ 0 & 0 & 1 \\ 0 & 1 & 0 \\ 0 & 1 & 1 \\ 1 & 0 & 0 \\ 1 & 0 & 1 \\ 1 & 1 & 1 \\ 1 & 1 & 0 \end{bmatrix} \tag{10.3.1}$$

There is more to this than just mathematical notation, however, because **T** has a deep *physical* property that I'll tell you about when we get to quantum logic in the final section of this chapter.

Another quite interesting logically reversible gate is the *Fredkin gate*, proposed in 1982 by Toffoli and Edward Fredkin (born in 1934, Fredkin is now at Carnegie Mellon University). As with the Toffoli gate, there are three inputs (A, B, and C) and three outputs (A, B′, and C′). Now, however, there is just *one* control input (A)—the value of A never changes from input to output—and the way the gate works is shown in the following truth table. When $A = 0$, B and C pass straight though to the outputs B' and C', respectively. That is, if $A = 0$ then $B' = B$ and $C' = C$. But if $A = 1$, then B and C are *swapped*, that is, if $A = 1$ then $B' = C$ and $C' = B$. The Fredkin gate is also called a *controlled-swap gate*.

A	B	C	A	B'	C'
0	0	0	0	0	0
0	0	1	0	0	1
0	1	0	0	1	0
0	1	1	0	1	1
1	0	0	1	0	0
1	0	1	1	1	0
1	1	0	1	0	1
1	1	1	1	1	1

From the truth table (and Karnaugh maps) we can write

$$B' = \bar{A}B\bar{C} + \bar{A}BC + A\bar{B}C + ABC = AC + \bar{A}B,$$

and

$$C' = \bar{A}\bar{B}C + \bar{A}BC + AB\bar{C} + ABC = AB + \bar{A}C.$$

These equations immediately tell us that if $B = 0$ then $B' = AC$ (we have an AND gate), and that if $B = 1$ and $C = 0$ then $B' = \bar{A}$ (we have a NOT gate). You can, of course, also see these relations from a direct inspection of the truth table, but I think the Boolean equations make them more obvious. (If $B = 0$ and $C = 1$ then $C' = \bar{A}$ and we also have a NOT.) Since we know we can build any Boolean function from ANDs and NOTs, then we know we can build any Boolean function from Fredkin gates.

As with the Toffoli gate, we can write the Fredkin gate in matrix notation as an 8 by 8 **F** "operating on" an 8 by 3 input to give an 8 by 3 output. That is, if

$$\mathbf{F} = \begin{bmatrix} 1 & 0 & 0 & 0 & 0 & 0 & 0 & 0 \\ 0 & 1 & 0 & 0 & 0 & 0 & 0 & 0 \\ 0 & 0 & 1 & 0 & 0 & 0 & 0 & 0 \\ 0 & 0 & 0 & 1 & 0 & 0 & 0 & 0 \\ 0 & 0 & 0 & 0 & 1 & 0 & 0 & 0 \\ 0 & 0 & 0 & 0 & 0 & 0 & 1 & 0 \\ 0 & 0 & 0 & 0 & 0 & 1 & 0 & 0 \\ 0 & 0 & 0 & 0 & 0 & 0 & 0 & 1 \end{bmatrix}$$

then

$$\mathbf{F}\begin{bmatrix} 0 & 0 & 0 \\ 0 & 0 & 1 \\ 0 & 1 & 0 \\ 0 & 1 & 1 \\ 1 & 0 & 0 \\ 1 & 0 & 1 \\ 1 & 1 & 0 \\ 1 & 1 & 1 \end{bmatrix} = \begin{bmatrix} 0 & 0 & 0 \\ 0 & 0 & 1 \\ 0 & 1 & 0 \\ 0 & 1 & 1 \\ 1 & 0 & 0 \\ 1 & 1 & 0 \\ 1 & 0 & 1 \\ 1 & 1 & 1 \end{bmatrix} \qquad (10.3.2)$$

Like **T**, **F** has the deep physical property I alluded to earlier that I'll discuss in the final section of this chapter.

A final observation: an important property of the Fredkin gate, a property not shared by the Toffoli gate, is the *preservation of parity*. That is, the outputs of a Fredkin gate always have the same number of 1s and 0s as do the inputs.[6]

10.4 THERMODYNAMICS OF LOGIC

The reason for our interest in logically reversible computation becomes clear once we ask the following question: for the logically *irreversible* gates, where does the destroyed information "go"? It appears as heat! An implicit recognition of this can be found as long ago as 1929, in an important thermodynamics paper by the Hungarian physicist Leo Szilard (1898–1964).[7] The explicit tying together of information, energy, and computation in analysts' minds is, however, almost certainly due to a remark made by the Institute for Advanced Study mathematician John von Neumann (1903–1957) in a December 1949 lecture at the University of Illinois.[8] In that lecture he asserted that the minimum energy E_{min} associated with manipulating a bit to be $k\text{T}\ln(2)$ joules (J), where T is the temperature on the Kelvin scale and k is Boltzmann's constant ($k = 1.38 \cdot 10^{-23} \frac{\text{J}}{\text{K}}$).[9] (Power is energy per unit time and so, just to keep the scale of this in mind, $1 \frac{\text{joule}}{\text{second}} = 1 \frac{\text{J}}{\text{s}} = 1$ watt) At "room temperature," that is, at $T = 300\,K$, this minimum energy is $2.87 \cdot 10^{-21}$ J, a *very* tiny amount of energy. As a comparison, every time *one* of the neurons in your brain "fires" (see the "majority logic" discussion in Chapter 6), the energy involved is a *hundred billion* (10^{11}) times larger than von Neumann's E_{min}![10]

The existence of logically reversible computation, which doesn't destroy information and so avoids the fundamental cause of heat, makes it plausible that it might be possible to build computers that could operate *below* von Neumann's E_{min}. This is of great practical interest because, while E_{min} may be very tiny, in a modern VLSI (Very Large-Scale Integrated) computer circuit chip running at a high clock rate, the amount of heat energy produced can result in substantial power levels. The observation that computers get hot seems to be obvious, if not actually trivial. After all, computers run on electricity, and every other electrical gadget we are familiar with (toasters, vacuum cleaners, television sets, light bulbs) gets hot. Why should computers be any different?

To be specific, imagine we have a VLSI chip with N transistors on it, operating at a clock frequency of f.[11] The workhorse technology in VLSI is—and has been for the last 20 years—CMOS (Complementary Metal-Oxide Semiconductors). I promised at the very start of the book that there would be no electronics required, and so I am not going to delve into the mathematical physics of a CMOS transistor. For our purposes here, we can simply imagine a CMOS transistor as a capacitor C that has either been charged to V volts (logic 1) or has been discharged to the ground voltage of zero volts (logic 0).

If we imagine C being charged to V volts by connecting C to a source voltage V through a path of resistance R, then it is a simple problem in first-year calculus and differential equations to show (which I am not going to do here) that the energy *stored* in C is $\frac{1}{2}CV^2$. Also, the charging current from the source that deposits that energy in C *dissipates* an *additional* $\frac{1}{2}CV^2$ of heat energy in R (notice that this dissipated energy is independent of R). So, it takes $\frac{1}{2}CV^2$ of energy to store a bit in a CMOS transistor. Sometime later, when that CMOS transistor is reset to logic 0, the stored energy is converted to more heat when the capacitor dumps its stored charge (as a current through some resistance path) to ground. So, we have $\frac{1}{2}CV^2$ of stored energy dissipated when the previously stored bit is destroyed. The total dissipated energy is therefore CV^2 when a CMOS transistor cycles from 0 to 1 to 0.

For a "typical" CMOS transistor, $C = 5$ femtofarads (that is, $C = 5 \cdot 10^{-15}$ farads). This value might vary in different CMOS devices, one way or the other, by a factor of 2, 5, or even more, but our final result will be "typical." If V is our usual 5 volts, then to place a 1 bit in our CMOS transistor requires the stored energy

$$\frac{1}{2} \cdot 5 \cdot 10^{-15} \cdot (5)^2 \,\mathrm{J} = 6.25 \cdot 10^{-14} \,\mathrm{J}.$$

To store this energy in C requires the same amount in *dissipated* energy, as mentioned above. The same amount of energy is also dissipated as heat when the stored bit is destroyed by dumping C's charge to ground. So, a total of $1.25 \cdot 10^{-13}$ J is dissipated as heat every time a CMOS transistor cycles from 0 to 1 to 0. This is nearly 22 *million times larger*(!) than E_{min}, and so we are today a long, *long* way from von Neumann's limit.

Now, suppose, on average, that at any given clock cycle $\frac{1}{2}N$ of the transistors on our VLSI chip are each storing a bit, and $\frac{1}{2}N$ of the transistors are each dumping their charge. That means with every clock cycle the total amount of dissipated energy is given by

$$N\frac{1}{2}CV^2 = 6.25 \cdot 10^{-14} N\,\text{J}.$$

Since the chip runs at clock frequency f, then the chip's total heat energy per second is $6.25 \cdot 10^{-14} Nf$ J or, with $N = 10^7$ transistors, this is $6.25 \cdot 10^{-7} f$ J. If the clock rate is $f = 1$ GHz, a pretty ordinary value today, then our chip dissipates energy at the rate of $6.25 \cdot 10^{-7} \cdot 10^9 \frac{\text{J}}{\text{s}} =$ 625 watts. This chip is going to need a *lot* of cooling to keep it from melting! This calculation makes the cause of the curious medical condition called "lap burn"—commonly suffered by airline passengers balancing their laptop computers on their knees while sitting in airport lounges—easy to understand.

One obvious way to reduce such a large power level is to simply reduce the spread in logic values. For example, if we use 1 volt for logic 1, rather than 5 volts, then we reduce the rate of energy lost as heat by a factor of $\left(\frac{5}{1}\right)^2 = 25$, which brings that 625 watts down to a more reasonable 25 watts. Another approach to achieving power reduction is to reduce the clock frequency. If we reduce f by a factor of 10, for example, then the power level drops by the same factor. (This works the other way, too—the power level *increases* as we increase f—which is why the use of overclocking to speed up a digital machine can be risky.) Of course, our computer now runs one-tenth as fast as before, and so this seems to be counterproductive. But, not to make too awful a pun, *not so fast*!

The famous *Moore's law* (named after Gordon Moore, the cofounder of Intel, who stated it in the mid-1960s) says that the packing density of transistors on a chip has historically grown exponentially with time; that is, the packing density has a constant doubling time (which is just about two years). So, if we wait just 3.33 doubling times ($2^{3.33} \approx 10$), that is, 7 years, we should be able to pack ten of our original chips into the same space as today. And if we were clever enough to discover how to use that factor of 10 *increase* in transistor count to balance that speed *decrease* by a factor of 10, then we could do the same computation (as a so-called *parallel computation*) in the same time as it takes today with our original chip. Well, okay, you say to that, but what you might also ask is: what have we really gained? After all, even though each of our densely packed ten chips is running at one-tenth the power level, there are ten chips where there used to be one, and so we seem to have the same total power level as before.

It turns out, surprisingly I think, that's not the case. Since the transistor packing density has *increased*, then the dimensions of each transistor must *decrease*, and that means the energy per transistor 0-to-1-to-0 cycle decreases,[12] and that means we'll actually need less total energy to do the same computation in the same time as today. That is, there will be a decrease in the total power level.

There is a limit to how far we can go with all this, however, as it will eventually become impossible to distinguish the energy of the 0-bit state of the transistor from the energy of the 1-bit state. If we use the von Neumann energy of $E_{\min}$ at $T = 300\,\text{K}$ for the 1-bit state and zero for the 0-bit state (and ignore all other difficulties!), then the famous *uncertainty principle* from quantum mechanics tells us that, if Δt is the time it takes to determine which energy state the transistor is in, then

$$\Delta t\, E_{\min} \geq \frac{h}{2}\pi$$

where h is Planck's constant ($6.63 \cdot 10^{-34}\,\text{J} \cdot \text{s}$). That is,

$$\Delta t \geq \frac{h}{2\pi E_{\min}}.$$

Since a clock period must be at least Δt, we have an upper bound on the clock frequency f as

$$f \leq \frac{1}{\Delta t} = \frac{2\pi E_{\text{min}}}{h} = \frac{2\pi \cdot 2.87 \cdot 10^{-21}}{6.63 \cdot 10^{-34}} s^{-1} = 27{,}200\,\text{GHz}.$$

Again, we are today a long, *long* way from such an incredible clock frequency!

10.5 A PEEK INTO THE TWILIGHT ZONE: QUANTUM COMPUTERS

In the final two sections of this book I'll sketch ideas from quantum mechanics that will make the fantastic calculations of the previous section look like mere childplay. To start, let me ask you the following question: if I give you a positive integer N, can you factor N? For example, what are all the factors of $N = 147{,}573{,}952{,}589{,}676{,}412{,}927$? In fact, there are two,[13] and one way to determine the answer for an arbitrary N is to simply divide N by all the integers from 2 to the largest integer no larger than $\sqrt{N}$ and check each division for a remainder. If a divisor doesn't generate a remainder, then the divisor is a factor of N. If a divisor does generate a remainder, that divisor is not a factor of N. If *all* possible divisors generate remainders, then N is prime (that is, N is not composite). This algorithm does work, but a mathematician would say it isn't *efficient* because it doesn't execute in what is called *polynomial time*.

The concept of an efficient algorithm is central to the theory of computing. An algorithm is *efficient* if, with input N, the time required to run to completion is bounded from above by some polynomial function of the size of N. The *size* of N is defined to be the number of bits needed to describe N, that is, $\log_2(N)$. So, just for example, if the time to execute for an algorithm is proportional to $3\,\{\log_2(N)\}^{10} + 7\{\log_2(N)\}^3$, then that algorithm would be declared to be efficient. If, on the other hand, the run-time increases faster than *any* finite power of $\log_2(N)$, then the algorithm would not be efficient. For the case of our factoring algorithm, its operation is essentially that of $\sqrt{N}-1$ divisions, which is essentially $\sqrt{N}$ divisions for any reasonably large N. That is, the run-time increases as

$$\sqrt{N}=N^{1/2}=2^{\log_2(N^{\frac{1}{2}})}=2^{\frac{1}{2}\log_2(N)}$$

which is an *exponentially* increasing function in $\log_2(N)$. This exponential increases faster than any polynomial with a finite maximum power term, and so, while our factoring algorithm is simple to understand, it is not efficient.

Being polynomial time doesn't mean a problem is trivial. An example of a challenging polynomial time problem is the *dating problem*. Given two lists of names, one of n men and one of n women, with the first list giving for each man the names of all the women he would be willing to date, and the second list giving for each woman the names of all the men she would be willing to date, is it possible to match each man and woman with an acceptable date? There *is* an efficient algorithm for doing that, but it may not be immediately obvious—I'll let *you* think about it!

Now, consider the reverse to the factoring problem. If I claim N is the product of two primes p and q, the values of which I tell you, can you check my claim? Sure you can, simply by multiplying p and q and seeing if you get N. The ease of multiplication and the difficulty of factoring is the key idea behind the supposed "unbreakability" of the now commonly used public-key cryptosystems. It is interesting to note that the importance of being able to factor was evident long before cryptosystems, with the great German mathematician Karl Gauss (1777–1855) writing in his 1801 masterpiece *Disquisitiones Arithmeticae*: "The problem of distinguishing prime numbers from composite numbers and of resolving the latter into their prime factors is known to be one of the most important and useful in arithmetic."

To give you a specific illustration of what a difficult problem factoring is, imagine you have a computer that can execute a million instructions per second (MIPS). If the computer runs day and night for a year, it will execute a total of $31.5\cdot10^{12}$ instructions $=31.5\,trillion$ instructions, an impressively large number computer scientists call a *MIPS-year*. In 1994 a 129 digit N formed by multiplying two large primes p and q (and used in one popular public-key cryptosystem) required 5,000 MIPS-years of computer effort to discover p and q. So, such systems *are* actually breakable, but only with the expenditure of enormous computational resources. And if someone, someday *does* factor the N in some cryptosystem, one could just pick larger p and q and generate a new N with *lots more* digits.

All the above comments are valid for what are called "classical" computers, computers made from the irreversible logic gates and flip-flops (made, in turn, from either relays or silicon transistors) we have discussed all through this book. But what if there were a new kind of computer, one with computational powers beyond the classical? What if we could make a "quantum computer"? Then it "theoretically" appears (on paper, anyway) that one could do almost supernaturally better with the factoring problem. For example, in 1994 Peter Shor (born 1959), a mathematician then at AT&T Bell Labs and now on the faculty at MIT, showed how such a computer (assuming it could ever be constructed) could efficiently, in polynomial time, factor an arbitrary N.[14] That naturally got the immediate attention of national security agencies all around the world who hope to keep their secret messages secret (and to break the codes of their rivals), as well as all the civilian users of public-key cryptosystems (credit card issuers and banks, for example).

As Shor himself amusingly expressed the implications of his discovery (in rhyme, no less):

> If computers that you build are quantum,
> Then spies everywhere will all want 'em.
> Our codes will all fail,
> And they'll read our email,
> Till we get crypto that's quantum and daunt 'em.[15]

A completely different problem, that of searching a data file that is unstructured with respect to a search parameter (such as, for example, looking for the name of a person when all you have is their telephone number), can also be done faster on a quantum computer as compared to a classical computer. Suppose the telephone book has N names. On a classical computer a random search is the best you can do, so, in fact, you might just as well start at the beginning and proceed alphabetically. That means you will have to examine at least 1, and perhaps as many as all N, entries. On average, your search will take $\frac{1}{2}N$ attempts. In 1996, however, Lov Grover (born 1961) at Bell Labs found a quantum search algorithm that requires only $\sqrt{N}$ attempts. So, if $N = 1{,}000{,}000$ names, then on a classical computer you can expect to take 500,000 attempts to complete your search, while Grover's quantum computer will finish in just 1,000 attempts.

Okay, what's a "quantum computer"? This is not an easy question to answer, at the level of this book, but I think I can at least give you a feeling for why a quantum computer could potentially perform computations in polynomial time, computations that are beyond a classical computer's abilities *to do efficiently*. Don't, however, fall into the fallacy of thinking this means a quantum computer could do *anything*. Even a quantum computer couldn't solve the halting problem—see note 2 again. I put the efficiency caveat in because it is possible to *simulate* any quantum computer on a classical computer, and so a classical computer could, *given enough time*, do anything a quantum computer can do—the rub is that the classical computer may take a vastly longer time to do it! A classical computer is to a quantum computer as a Turing machine is to a classical computer (although that comparison might give the classical computer more than it deserves).

Before going any further I should tell you that not everybody is convinced quantum computation is the ultimate pot of gold at the end of the computer rainbow. One such healthy skeptic is Scott Aaronson (born 1981), a professor of electrical engineering and computer science at MIT. He has described problems that would remain beyond the power of even a quantum computer to solve efficiently.[16] One such problem is the following: given an arbitrary planar map, can it be colored with just three colors so that no two countries that share a common border have the same color? Since the 1976 computer-based proof that, in general, it requires *four* colors to color a planar map, we know that there are maps for which three colors would not be sufficient. But, nevertheless, there *are* particular maps for which three colors are enough. There is, alas, no efficient algorithm known that can distinguish between three-color and four-color maps, and so having a quantum computer available would be of no help.

And even when a quantum algorithm is known, it may not result in a polynomial time computation. Grover's search algorithm is such a case because, while $\sqrt{N}$ is indeed faster than N, we know from our earlier discussion of the factoring problem that $\sqrt{N}$ is still exponential and not polynomial.

A quantum computer makes direct use of one of numerous available quantum mechanical phenomena to manipulate information. Classical computers manipulate bits that are each in one of two definite states (0 or 1)—for example, a closed (or open) relay contact, or stored charge (or not) in a CMOS transistor. A quantum computer also manipulates bits—called *quantum bits* or *qubits*—but qubits are *not* restricted to being in one of just two states. For a single qubit there are indeed two pure states, written as $|0\rangle$ and $|1\rangle$, where $|0\rangle$ and $|1\rangle$ are short-hand for the column vectors

$$\begin{bmatrix} 1 \\ 0 \end{bmatrix} \text{and} \begin{bmatrix} 0 \\ 1 \end{bmatrix},$$

respectively. These two pure quantum states are analogous to the two classical bit states, but in quantum mechanics each qubit most generally exists as a linear combination or *superposition* of all the pure states. This is not intuitive, so don't worry if it seems odd, because it's odd to everybody! Indeed, even the great Einstein found quantum mechanics hard to fully accept, and he left this world thinking the theory to be at best incomplete (his word) in its description of reality. In the next section I'll show you why he felt that way.

In the early 1960s, just a few years after Einstein's death, the Irish physicist John Bell (1928–1990) developed a mathematical theory that showed how in principle one could perform experimental tests to determine the truth or not of quantum mechanics. In the early 1980s those experiments were finally definitively performed, and the results fully supported quantum mechanics. The superposition of pure states idea—odd as it is—is accepted today because it results in the prediction of effects that are *verifiable* effects. Quantum mechanics is the most successful scientific theory ever developed, with *every one* of its predictions that have been tested by experiment being confirmed.

If we are talking about a single qubit system named ψ, then the so-called *state-vector* of that system is written as

$$|\psi\rangle = c_1\,|0\rangle + c_2|1\rangle = c_1\begin{bmatrix}1\\0\end{bmatrix} + c_2\begin{bmatrix}0\\1\end{bmatrix} = \begin{bmatrix}c_1\\0\end{bmatrix} + \begin{bmatrix}0\\c_2\end{bmatrix} = \begin{bmatrix}c_1\\c_2\end{bmatrix}, \quad (10.5.1)$$

where c_1 and c_2 are complex numbers (about which I'll tell you more in just a moment). The physical significance of the pure states of an individual qubit could come from any number of possibilities; the polarization of a photon is a common example.[17] Thus, one polarization state would be $|0\rangle$ and the other polarization state would be $|1\rangle$. Whether an orbital electron in an atom is in its ground state or in a higher-energy "excited" state (induced, for example, by shining laser light on the atom) is another possibility, and yet another is the *spin* of atomic nuclei (that measures something called the *magnetic moment* of the nuclei). Spin is a quantized quantity, and it can be switched from one allowed value to another by illuminating the nuclei with microwave radiation of the proper frequency (called the *resonant* frequency). This process, called *nuclear magnetic resonance* (or NMR), was the quantum mechanical phenomenon used, in late 2001, to factor 15 using Shor's algorithm.[18] A fourth possible technology is the so-called *ion trap*, in which positive ions (atoms missing orbital electrons) that act like tiny magnets are suspended in a vacuum by electric fields produced by nearby tiny electrodes. The magnetic states of the ions can be manipulated by targeting them with ultrathin, threadlike laser beams.

The state-vector $|\psi\rangle$ is in its superposition form until we measure it, and then the state-vector collapses from ghostly superposition to one of its definite pure states. This collapse is probabilistic, a feature of quantum mechanics that gives it its nonintuitive behavior.[19] The probability a single qubit system collapses to $|0\rangle$ is $|c_1|^2$ (that is, the product of c_1 and c_1^* where c_1^* is the complex conjugate of c_1), and the probability it collapses to $|1\rangle$ is $|c_2|^2$. Since the observed qubit has to be one or the other of the pure states, we thus have the so-called *probability constraint* $|c_1|^2 + |c_2|^2 = 1$. Why is all this so? Nobody knows. Does that bother you? Well, why is nothing able to go faster than the speed of light? Nobody knows that, either. *That* probably doesn't bother you, but

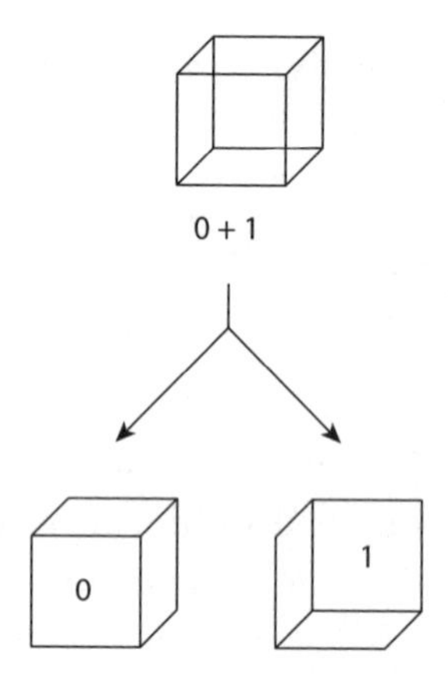

Figure 10.5.1. An optical illusion illustrating both state superposition and measurement collapse.

that's only because you've heard it all your life and are just used to it. People will get used to quantum mechanics, too, if quantum computers become commonplace.

The ideas of *superposition of states* and *state-vector collapse* can be illustrated by the well-known optical illusion shown in Figure 10.5.1. When you look at the two-dimensional rendition of a three-dimensional cube, you can alternately "see" two different spatial orientations for the cube, which are labeled 0 and 1. When we see either specific orientation, we have "collapsed" the cube into one or the other "measured" state, but until we do that the cube is simultaneously in both states.

If ψ is a two-qubit system we would write the four possible pure states as $|00\rangle$, $|01\rangle$, $|10\rangle$, and $|11\rangle$, and the state-vector of ψ as the superposition

$$|\psi\rangle = c_1|00\rangle + c_2|01\rangle + c_3|10\rangle + c_4|11\rangle,$$

where $|c_1|^2 + |c_2|^2 + |c_3|^2 + |c_4|^2 = 1$. In general, a system of n qubits has a state-vector of the form

$$|\psi\rangle = c_1|00\ldots00\rangle + c_2|00\ldots01\rangle + c_3|00\ldots10\rangle + \cdots + c_{2^n}|11\ldots11\rangle.$$

You can probably now see our first astonishing departure of a quantum computer from a classical computer. In a classical computer, the description of a collection of n bits requires just n discrete numbers (each either 0 or 1). In a quantum computer, the description of the state-vector of n qubits requires 2^n numbers, each of which takes its value from a lot more than just two possibilities, with that huge collection of numbers obeying the probability constraint

$$\sum_{k=1}^{2^n} |c_k|^2 = 1.$$

An n qubit quantum system obviously has a lot more information packed into it than does an n bit classical system (for $n = 500$, for example, $2^{500} \approx 3.24 \cdot 10^{150}$, a number larger than the total number of elementary particles in the entire universe).

How Nature keeps track of all those numbers is another of the deep mysteries of quantum mechanics, but it certainly makes plausible the claim that processing qubits is far different from processing classical bits. It is this phenomenal ability to process an enormous number of numbers that gives a quantum computer its power. When a quantum computer processes an n qubit system, it is simultaneously operating on all possible 2^n pure states, not just one as does a classical computer. That is, a quantum computer actually follows a multitude of superimposed, parallel computational paths (a process called *quantum interference*) to arrive at its final result. Such massive parallelism is the fundamental origin of the quantum computer's phenomenal speed.

A class of famous, easy-to-understand problems (quite different from the factoring and search problems) that could also benefit from the massive parallelism of a quantum computer are the so-called *knapsack problems*. They are old problems, dating back at least to 1897, but they have modern applications, including cryptography. One knapsack problem asks the following question: given a stretchable knapsack that can hold a maximum weight of W, and n objects of individual weights $w_1, w_2, \ldots, w_n$, is there a subset of these objects that can be put into the knapsack that just achieves the knapsack's weight capacity? That is, is there a solution for the x_i in the equation

$$W = w_1x_1 + w_2x_2 + \cdots + w_nx_n,$$

where $x_i = 0$ or 1 for $1 \le i \le n$?

For example, $W = 22$ and $w_1 = 3$, $w_2 = 7$, $w_3 = 9$, $w_4 = 11$, and $w_5 = 20$ has no solution, while $W = 27$ for the same w_i has two solutions ($x_1 = x_5 = 0$ and $x_2 = x_3 = x_4 = 1$, and $x_1 = x_3 = x_4 = 0$ and $x_2 = x_5 = 1$). For certain special cases of the w_i there are algorithms for quickly finding solutions (if they exist), but in the general case the only way to attack the problem is the elementary one of trying, one after the other, each of the 2^n possibilities for the n x_i. For a present-day classical computer this becomes more and more impractical once n grows beyond 100 or so. With a quantum computer, all 2^n possibilities could conceivably be examined simultaneously.

The probabilistic nature of quantum interference and state-vector collapse means that the final output of a quantum computer is probabilistic and may, in fact, actually be wrong! Shor's factoring algorithm, in particular, has that characteristic. It is, however, *not* a particularly troublesome problem, an observation that usually puzzles people when they first hear it. To see this, suppose we have the output from a quantum computer execution of Shor's algorithm. That is, it has declared F to be a factor of N. It's duck soup to simply check F by dividing N by F and then see if we have a remainder. If there isn't one, then F is correct. If there is a remainder, then F is wrong and so just run Shor's algorithm again. If the probability is $1 - \epsilon$ that the algorithm gives a valid F (that is, ϵ is the probability that F is wrong), then, if we run the algorithm k times, the probability that we get wrong answers all k times is ϵ^k, which means we'll get a correct answer at least once with probability $1 - \epsilon^k$. And all we need is to get a correct answer *once*. As long as $\epsilon < 1$, then we can make the probability of a correct answer as close to 1 as we wish just by making k big enough. We can afford to do all this because Shor's algorithm is efficient.

This sort of nonclassical behavior may be what Australian science fiction writer Greg Egan had in mind in his 1992 novel *Quarantine: A Novel of Quantum Catastrophe* when at one point he has his protagonist muse, "Computerized information is as evanescent as the quantum vacuum, with virtual truths and falsehoods endlessly popping in and out of existence."

10.6 QUANTUM LOGIC—AND TIME TRAVEL, TOO!

How do we process qubits? We do that with what are called *quantum logic gates*. If we look at just the simplest such gate, the quantum inverter for a single qubit, we'll see that even in that most elementary case there are profound differences between the quantum gate and its classical analogue (the NOT). The mathematics of the quantum inverter is actually not difficult, which perhaps makes the surprising results even more surprising!

It seems reasonable to argue that, given the state-vector $|\psi\rangle$ in (10.5.1), a quantum inverter would simply swap the $|0\rangle$ and $|1\rangle$ pure states. That is, after inversion we should have the new state-vector:

$$|\psi'\rangle = c_1|1\rangle + c_2|0\rangle. \tag{10.6.1}$$

It may not be immediately apparent from (10.5.1) and (10.6.1), but this behavior has a particularly nice mathematical form if we express the two state-vectors in matrix form. That is, if we write the pure states $|0\rangle$ and $|1\rangle$ in their column vector form, then the state-vector in (10.6.1), at the output of the inverter, is

$$|\psi'\rangle = \begin{bmatrix} c_2 \\ c_1 \end{bmatrix}. \tag{10.6.2}$$

If we represent the quantum inverter gate by the symbol **N**, then **N** "operating on" $|\psi\rangle$ should give $|\psi'\rangle$. That is,

$$\mathbf{N}\begin{bmatrix} c_1 \\ c_2 \end{bmatrix} = \begin{bmatrix} c_2 \\ c_1 \end{bmatrix}. \tag{10.6.3}$$

What "operating on" a two-element column vector gives another two-element column vector? A 2 by 2 matrix! That is, if we write

$$\mathbf{N}\begin{bmatrix} c_1 \\ c_2 \end{bmatrix} = \begin{bmatrix} a & b \\ d & e \end{bmatrix}\begin{bmatrix} c_1 \\ c_2 \end{bmatrix} = \begin{bmatrix} c_2 \\ c_1 \end{bmatrix}.$$

then performing the matrix multiplication gives us the two equations

$$ac_1 + bc_2 = c_2$$
$$dc_1 + ec_2 = c_1,$$

which, by inspection, say $a = 0$, $b = 1$, and $d = 1$, $e = 0$. That is, the quantum logic inverter gate is, *mathematically*, the matrix

$$\mathbf{N} = \begin{bmatrix} 0 & 1 \\ 1 & 0 \end{bmatrix}. \tag{10.6.4}$$

Our result in (10.6.4) makes sense, too, when you ask yourself the question: what should happen to $|\psi\rangle$ if we run it through *two* quantum inverters in series? The answer seems clear: you should get $|\psi\rangle$ back. Do we? Yes, because

$$\mathbf{N}\{\mathbf{N}|\psi\rangle\} = \mathbf{N}\left\{\mathbf{N}\begin{bmatrix} c_1 \\ c_2 \end{bmatrix}\right\} = \mathbf{N}\begin{bmatrix} c_2 \\ c_1 \end{bmatrix} = \begin{bmatrix} 0 & 1 \\ 1 & 0 \end{bmatrix}\begin{bmatrix} c_2 \\ c_1 \end{bmatrix} = \begin{bmatrix} c_1 \\ c_2 \end{bmatrix} = |\psi\rangle$$

And, in fact,

$$\mathbf{NN} = \begin{bmatrix} 0 & 1 \\ 1 & 0 \end{bmatrix}\begin{bmatrix} 0 & 1 \\ 1 & 0 \end{bmatrix} = \begin{bmatrix} 1 & 0 \\ 0 & 1 \end{bmatrix} = \mathbf{I}$$

where $\mathbf{I}$ is the 2 by 2 identity matrix and (of course!) $\mathbf{I}\,|\psi\rangle = |\psi\rangle$.

Unlike the classical NOT, which is the only nontrivial logical function of a single bit, the quantum inverter is just one of many possible quantum operations that can be usefully performed on a single qubit. This is yet another indication of how quantum computers are quite different from classical ones. To see this, let's write a general 2 by 2 matrix $\mathbf{M}$ and let it operate on the state-vector $|\psi\rangle$ to give

$$\begin{aligned} \mathbf{M}\,|\psi\rangle = |\psi'\rangle &= \begin{bmatrix} a & b \\ d & e \end{bmatrix}\begin{bmatrix} c_1 \\ c_2 \end{bmatrix} = \begin{bmatrix} ac_1 + bc_2 \\ dc_1 + ec_2 \end{bmatrix} \\ &= (ac_1 + bc_2)\,|0\rangle + (dc_1 + ec_2)\,|1\rangle. \end{aligned}$$

Now, let's write $\mathbf{M}^\dagger$ as the *adjoint* of M, which means $\mathbf{M}^\dagger$ is the *conjugated transpose* of $\mathbf{M}$. That is,

$$\mathbf{M}^\dagger = \begin{bmatrix} a^* & d^* \\ b^* & e^* \end{bmatrix}$$

And suppose further that we require $\mathbf{M}^\dagger\mathbf{M} = \mathbf{I}$. That is,

$$\begin{bmatrix} a^* & d^* \\ b^* & e^* \end{bmatrix}\begin{bmatrix} a & b \\ d & e \end{bmatrix} = \begin{bmatrix} 1 & 0 \\ 0 & 1 \end{bmatrix}$$

Then,

$$\begin{aligned} |a|^2 + |d|^2 &= 1 \\ ab^* + e^*d &= 0 \\ a^*b + d^*e &= 0 \\ |b|^2 + |e|^2 &= 1. \end{aligned} \tag{10.6.5}$$

If $\mathbf{M}$ has all these mathematical properties, then it has a wonderful *physical* property as well, which I'll explain next.

Since all state-vectors must satisfy the probability constraint, then in particular $|\psi'\rangle$ must as well, and so we have

$$(ac_1 + bc_2)(ac_1 + bc_2)^* + (dc_1 + ec_2)(dc_1 + ec_2)^* = 1,$$

or, as the conjugate of a sum (product) is the sum (product) of the conjugates,

$$(ac_1 + bc_2)(a^*c_1^* + b^*c_2^*) + (dc_1 + ec_2)(d^*c_1^* + e^*c_2^*) = 1,$$

or,

$$\begin{aligned} &|a|^2|c_1|^2 + ba^*c_2c_1^* + ab^*c_1c_2^* + |b|^2|c_2|^2 \\ &+ |d|^2|c_1|^2 + de^*c_1c_2^* + ed^*c_2c_1^* + |e|^2|c_2|^2 = 1 \end{aligned}$$

or,

$$\{|a|^2+|d|^2\}|c_1|^2+\{|b|^2+|e|^2\}|c_2|^2$$
$$+\{a^*b+d^*e\}c_2c_1^*+\{ab^*+de^*\}c_1c_2^*=1.$$

But, from (10.6.5) we see that the last two terms on the left are zero, and that the coefficients of $|c_1|^2$ and $|c_2|^2$ are each 1, and so we are left with

$$|c_1|^2+|c_2|^2=1,$$

which is certainly true as it is the probability constraint on the original state-vector $|\psi\rangle$.

That is, if $\mathbf{M}$ is *any* 2 by 2 matrix with an adjoint such that $\mathbf{M}^\dagger\mathbf{M}=\mathbf{I}$, then the probability constraint will automatically be satisfied by **M**'s output state-vector $\mathbf{M}|\psi\rangle=|\psi'\rangle$. Quantum physicists say such **M** matrices *preserve probability* and call them *unitary*. Our quantum inverter gate **N** is unitary, but it is not the only such gate. Another particularly interesting one is

$$\mathbf{H}=\frac{1}{\sqrt{2}}\begin{bmatrix}1 & 1\\ 1 & -1\end{bmatrix}.$$

Notice that **H** is its own adjoint (**H** is *self-adjoint*); mathematicians say **H** is *Hermitian*, after the French mathematician Charles Hermite (1822–1901). But that's not the reason we use the symbol **H**—I'll tell you why in just a moment. If **H** operates on $|0\rangle$ and $|1\rangle$, then

$$\mathbf{H}|0\rangle=\frac{1}{\sqrt{2}}\begin{bmatrix}1 & 1\\ 1 & -1\end{bmatrix}\begin{bmatrix}1\\ 0\end{bmatrix}=\frac{1}{\sqrt{2}}\begin{bmatrix}1\\ 1\end{bmatrix} \tag{10.6.6}$$
$$=\frac{1}{\sqrt{2}}\begin{bmatrix}1\\ 0\end{bmatrix}+\frac{1}{\sqrt{2}}\begin{bmatrix}0\\ 1\end{bmatrix}=\frac{|0\rangle+|1\rangle}{\sqrt{2}}$$

and

$$\mathbf{H}\,|1\rangle = \frac{1}{\sqrt{2}}\begin{bmatrix}1 & 1\\ 1 & -1\end{bmatrix}\begin{bmatrix}0\\ 1\end{bmatrix} = \frac{1}{\sqrt{2}}\begin{bmatrix}1\\ -1\end{bmatrix}$$

$$= \frac{1}{\sqrt{2}}\begin{bmatrix}1\\ 0\end{bmatrix} - \frac{1}{\sqrt{2}}\begin{bmatrix}0\\ 1\end{bmatrix} = \frac{|0\rangle - |1\rangle}{\sqrt{2}}. \qquad (10.6.7)$$

Since one could argue—very loosely!—that the right-hand side of (10.6.6) is "halfway" between turning $|0\rangle$ into $|1\rangle$, and the right-hand side of (10.6.7) is also "halfway" between turning $|1\rangle$ into $|0\rangle$, then **H** is "sorta like" a *square-root of NOT*![20] Some writers have even written the intriguing identity

$$\mathbf{H} = \sqrt{\mathbf{N}} = \sqrt{\text{NOT}},$$

but more conservative types generally call **H** the *Hadamard* quantum gate (after the French mathematician Jacques Hadamard (1865–1963)—*that's* where the **H** comes from!—who studied matrices with similar mathematical structures in the 19th century, long before their appearance in quantum mechanics). If **H** really was $\sqrt{\mathbf{N}}$ then we'd expect $\mathbf{H}^2 = \mathbf{N}$, but it's easy to confirm that $\mathbf{H}^2 = \mathbf{I} \neq \mathbf{N}$.

H is easily verified to be unitary, and in fact *all* quantum logic gates have unitary matrices. If you look back at **T** and **F** in Section 10.3, for example, you should be able to quickly verify that both the Toffoli and Fredkin quantum logic gates are unitary (that's the "deep physical property" I mentioned back there). And as a fun exercise, you should confirm for yourself that the matrix for the quantum controlled-NOT gate (the XOR in classical computers—see note 6 in Chapter 7) is also unitary. (See the hint at the end of the notes, p. 209.)

In the previous section I mentioned Einstein's unhappiness with quantum mechanics. Now, with the Hadamard quantum logic gate in hand, I can tell you why he felt that way. Imagine that we start with two individual ions, with both having the pure state-vector $|0\rangle$. One of these ions is then operated on by a Hadamard gate, as defined in

(10.6.6), giving the ion the output superposition state-vector

$$\frac{1}{\sqrt{2}}\,|0\rangle + \frac{1}{\sqrt{2}}\,|1\rangle.$$

This output is then used as the control input to a quantum CNOT gate, while the other input (the *controlled* input) is the other original ion with the pure state-vector $|0\rangle$. Since a control input in state 0 leaves the controlled input state unchanged, while a control input in state 1 flips the controlled input state, the CNOT output state-vector of the two-ion system is

$$\frac{1}{\sqrt{2}}\,|00\rangle + \frac{1}{\sqrt{2}}\,|11\rangle.$$

This means the two ions are now in the *same state* (either both 0 or both 1), but we don't know which of these two equally-likely possibilities it is since the output state-vector is a superposition state-vector. The two ions are said to have *entangled* states.

Now, here's where the puzzle that so bothered Einstein comes into play. Suppose Alice puts one of the ions in her suitcase, and Bob puts the other ion in his suitcase, and then both hop into separate rockets and fly away from each other with their suitcases until they are both in the orbit of Mars (on opposite sides of the Sun). That is, they are so far apart that a light-speed signal would take quite some time to travel from one to the other. Also in orbit around the Sun is a satellite with a radio transmitter, positioned so that it has direct line-of-sight paths of equal lengths to Alice and Bob. Once Alice and Bob are in Mars's orbit, the satellite transmits a radio pulse, which arrives at Alice and Bob at the same instant.

Upon receiving the radio pulse, Alice opens her suitcase, removes her ion, and measures the ion's state. This collapses the state-vector of the two-ion system, and Alice gets either 0 or 1 with equal probability. One minute after receiving the radio pulse (and so after Alice's measurement), Bob opens his suitcase and measures the state of his ion. Now there is nothing probabilistic about his result, because it must be the same as Alice's result (you'll recall that the two ions were *forced* to be in the same state). Since Alice could have gotten either a 0 or a 1, then somehow that random outcome showed up at Bob's location, in far less time than a light-speed signal would take to make the trip from Alice to Bob, to give the *certain*, *identical* result for Bob's measurement.

Some philosophers claim this situation violates special relativity's claim that "nothing can go faster than light," but that's not true and physicist Einstein knew that. Special relativity says mass-energy-information cannot go faster than light; but since Alice's measurement result was a random result and not under her control, she couldn't use the result to send information to Bob. Special relativity isn't bothered by this, and neither was Einstein. What bothered Einstein was how Alice's *random* result could instantly determine Bob's *certain* result, something Einstein called a "spooky action-at-a-distance." All this is still a matter of some controversy in physics, and is yet another indication of the extreme nonintuitive nature of quantum mechanics.

We could go on and on with quantum logic gate mathematics but of course the monster question lurking under the bed is: how do we actually *build* a quantum computer as opposed to just writing lots of matrix equations? Well, as I write there are only the most rudimentary ideas of how to do that. In the case of a photon quantum computer, where polarized photons provide the underlying quantum mechanism, the hardware would (not surprisingly) be optical in nature, utilizing mirrors, beamsplitters, interferometers, photodetectors, and such. Using such devices, the construction details for a Fredkin quantum logic gate were described in the literature as long ago as 1989.[21] A quantum computer built from such a building block would be very different from a classical computer built from silicon transistors on integrated circuit chips. A photon quantum computer would literally "do it with mirrors." (A few years later, in 1995, a controlled-NOT quantum logic gate using a nonphoton technology was actually demonstrated, and there have already been experiments done on building quantum logic gates on a chip using ion trap technology.[22]) And how to easily program a quantum computer as we can with a classical computer has not, I believe, received much (if any) attention at all.

None of these enormous practical issues has stopped some highly adventurous physicists from going to even more extreme heights of imagination. Suppose, goes their wildest speculations, just suppose—that we could combine a quantum computer with a time machine! Such a wondrous-squared machine (Professor Aaronson—see note 16—says such a combination would make ordinary quantum computers "look as pedestrian as vending machines") could send qubits into the past and so produce answers in zero time. With a time machine it would seem that you don't really need a quantum computer at all. Just let a classical computer grind away for however long it takes (perhaps years or decades or more) to finally get an answer, and then use the time machine to send the answer back in time to one nanosecond after you turned the computer on! Actually, matters aren't quite that simple, because of potential paradoxes. For example, suppose we turned the computer off one nanosecond after we get the answer from the far future. How then was the answer ever computed? Philosophers call this a *bilking paradox*.[23]

I am not joking with all this, and you can find such time travel possibilities discussed in the quite serious physics literature going back to 1991.[24] Even those whose enthusiasm for time traveling quantum computers is apparently unbounded know, however, that they have to show some restraint to avoid appearing merely goofy. As one writer ended his paper, "We would not be honest if we did not end this paper with the caveat that this work is at best a creature of eager speculation," and with the admission that "practical considerations are humorous at best."[25] Indeed.

Still, one has to be careful in pooh-poohing future scientific discoveries that might, no matter how unlikely they may seem, just barely be possible. The cautionary example of the Victorian scientist William Thomson (better known today as Lord Kelvin) should be kept in mind; as the nineteenth century became the twentieth, he declared there was nothing left in physics to be discovered, that there was no future for radio, that heavier-than-air flying machines were impossible, and that X-rays would prove to be a fraud. Those awful missteps ought to be a warning to all who indulge in predicting the future. So, having said that, let me now ignore my own warning.

I do think quantum computers will some day come to pass. As the authors of one paper (see note 14) so nicely put it, "The history of computer technology has involved a sequence of changes from one type of physical realization to another—from gears to relays to valves [that is, vacuum tubes] to transistors to integrated circuits and so on. The step to the molecular scale—the quantum level—will be next." That was written more than fifteen years ago, and we are still waiting, but I do think it will happen. On the other hand, I do not think we'll ever couple a quantum computer with a time machine, but I'll admit there is a very tiny chance I could be wrong (but even if I am wrong, I think it will be a very long time coming). So, that's what *I* think. What would Boole or Shannon have thought?

Shannon would surely have embraced quantum computation. How, indeed, could a brain like his have resisted the intellectual challenges? So would have Boole, once he had taken a crash course in modern physics and learned the jargon. As for the time travel part, however, well of course the only way we could ever know about that is if we could somehow travel back to their times and ask them. But if we could do that, well, what then would there be for them to doubt?

NOTES AND REFERENCES

1. P. J. Nahin, *Number-Crunching*, Princeton, 2011, p. 328.

2. The halting problem asks the following question: is it possible to write a computer program (in *any* language) that accepts as input *any* program (along with some arbitrary input) and then decides if that program (with that arbitrary input) will ever halt after being started? In *Number-Crunching* (pp. 330–333) I present the classic proof that it is logically impossible for such a halting-determination program to exist. There is no physics in that proof.

3. Charles Babbage (1791–1871), a British mathematician and computer inventor. He first designed the so-called *difference engine* and then the *analytical engine*, both to be built from gears, levers, ratchets, springs, nuts, and bolts. Babbage actually built neither one, but versions of the difference engine were constructed later by others. The design of mechanical computers can actually be traced back hundreds of years before Babbage: see Herman H. Goldstine, *The Computer from Pascal to von Neumann*, Princeton University Press, 1972.

4. Charles H. Bennett, "Notes on the History of Reversible Computation," *IBM Journal of Research and Development*, January 1988, pp. 16–23.

5. Charles H. Bennett and Rolf Landauer, "The Fundamental Limits of Computation," *Scientific American*, July 1985, pp. 48–56. The opening quotation to this chapter comes from this paper.

6. In *Feynman Lectures on Computation* (edited by A.J.G. Hey and R. W. Allen), Addison-Wesley, 1996, the American physicist Richard Feynman (1918–1988) says of this conservation property (p. 39): "Fredkin . . . demanded that not only must a gate be reversible, but the number of 1s and 0s should never change. *There is no good reason for this* [my emphasis], but he did it anyway." Contrary to Feynman, you'll see in the last section of this chapter that there can be, in fact, a very good reason for preserving parity.

7. Szilard's 1929 German-language paper, "On the Decrease of Entropy in a Thermodynamic System by the Intervention of Intelligent Beings," can be found in English translation in Harvey S. Leff and Andrew F. Rex, *Maxwell's Demon: Entropy, Information, Computing*, Princeton University Press, 1990, pp. 124–133. Szilard was a good friend of Einstein's (they jointly held a patent on a novel refrigerator—it had no moving parts!), and he helped write the first draft of the famous August 1939 letter sent by Einstein to President Roosevelt that initiated the American atomic bomb program in World War II.

8. See *Papers of John von Neumann on Computing and Computer Theory*, MIT Press, 1987, pp. 434–490.

9. You can find a derivation (calculus required!) of $E_{\min}$ in Feynman's *Lectures on Computation* (see note 6), pp. 137–146. Feynman's method is actually due to von Neumann, who described the approach (involving the energy required to compress an ideal gas consisting of just one molecule) in his 1932 book *Mathematical Foundations of Quantum Mechanics* (an English translation from the German was published by Princeton University Press in 1955; see pp. 370–372 of that book).

10. The 10^{11} factor mentioned in the text was calculated by von Neumann in the following way. He estimated the power level of a human brain to be 25 watts $(=25\frac{\text{J}}{\text{s}})$. If the brain has 10^{10} neurons, each firing ten times per second, then there are 10^{11} firings/second and so each firing must require an energy of $25 \cdot 10^{-11}\,\text{J} = 2.5 \cdot 10^{-10}\,\text{J}$. This energy is just about 10^{11} times larger than $E_{\min}$.

11. My illustration is inspired by Neil Gershenfeld, "Signal Entropy and the Thermodynamics of Computation," *IBM Systems Journal* 35, 1996, pp. 577–586.

12. See Carver A. Mead, "Scaling of MOS Technology to Submicrometer Feature Sizes," 1994, reprinted in Anthony J. G. Hey, *Feynman and Computation*, Perseus Books, 1999, pp. 93–115. Mead shows that if s is the characteristic feature size of a transistor, then the stored energy in C scales as $s^{2.75}$. To achieve a ten-fold increase in the number of transistors on a fixed-sized chip we therefore need to reduce the linear dimensions of a transistor by the factor of $\sqrt{10} = 3.16$ (that gives an area reduction by a factor of 10) and so the reduction in the required energy to store a bit is by a factor of $3.16^{2.75} = 23.7 > 10$. (If energy scaled as s^2, then there would be no gain; it's the extra 0.75 in the exponent that makes all the difference.)

13. $N = 147,573,952,589,676,412,927 = 2^{67} - 1$ was for many years speculated to be prime. The discovery of its prime factors (193,707,721 and 761,838,257,287) was a big event in 1903 mathematics. That year was long before the construction of even the crudest of electronic computers, of course, and the calculations were performed entirely by hand. The American mathematician who did that, Frank Cole (1861–1926), said it took him "twenty years of Sunday afternoons."

14. See Artur Ekert and Richard Jozsa, "Quantum Computation and Shor's Factoring Algorithm," *Reviews of Modern Physics*, July 1996, pp. 733–753. The only restrictions on Shor's algorithm are that neither p nor q can be 2 (the only even prime), and that p and q must be different. You can find a detailed discussion of the algorithm, at a math level similar to this book, in Oliver Morsch, *Quantum Bits and Quantum Secrets: How Quantum Physics Is Revolutionizing Codes and Computers*, Wiley-VCH, 2008. In 1981, the mathematician Manuel Blum (born 1938) showed how to use the difficulty of factoring to solve a seemingly impossible problem: two people remotely flipping a coin over a telephone circuit, so that each feels confident that the other isn't cheating. You can read how Blum (whose PhD advisor at MIT was Marvin Minsky, Shannon's inspiration for the "Ultimate Machine" discussed in Chapter 3) did this in Charles Vanden Eynden, "Flipping a Coin over the Telephone," *Mathematics Magazine*, June 1989, pp. 169–172.

15. Quoted from Michael A. Nielsen and Isaac L. Chuang, *Quantum Computation and Quantum Information*, Cambridge University Press, 2000, p. 216.

16. Scott Aaronson, "The Limits of Quantum Computers," *Scientific American*, March 2008, pp. 62–69.

17. To speak of the polarization of a single photon often strikes people as puzzling because we are used to thinking of *polarization* as being a property of electromagnetic radiation, a property determined by the spatial orientation of the electric field vector in a *wave* theory of light. Photons, on the other hand, are thought of as *particles*. How can a particle be polarized? This point is explained in the book by Gerard J. Milburn, *The Feynman Processor*, Perseus Books 1998, pp. 20–26. I'll let you look it up if you want to read the details, and I'll simply take it as fact that it *does* make sense to talk about polarization states for photons and that the state can be "measured" through the well-known use of optical filters.

18. Lieven M. K. Vandersypen et al., "Experimental Realization of Shor's Quantum Factoring Algorithm Using Nuclear Magnetic Resonance," *Nature*, December 20, 2001, pp. 883–887. The choice of the number $N = 15$ to factor was not arbitrary. It is the smallest possible value for $N = pq$, where p and q are both prime. Here's why. The first three primes are 2, 3, and 5. Since neither p nor q can be 2 (see note 14), and since $p = q$ is not allowed, then $p = 3$ and $q = 5$ are the smallest allowed values for the prime factors of N.

19. We don't want state-vector collapse to occur until the computation is completed, and the only way to achieve that is to keep a quantum computer *decoupled* from its surrounding environment. *Any* coupling is equivalent to observing the qubits, causing premature collapse, a phenomenon called *decoherence*. All realistic quantum computer technologies eventually succumb to this degradation. The trick is to get the computation done in a time less than the decoherence time (which varies from microseconds to minutes, depending on what quantum technology is used).The collapse of a measured (observed) quantum state-vector is the origin of the famous paradox of *Schrödinger's dead/alive cat*. The mystery of state-vector collapse with observation is what was behind Einstein's famous skeptical question: "Do you really believe the Moon exists only when you look at it?"

20. This curious name was probably motivated by the following geometric imagery. The numbers -1 and $+1$ are vectors on the horizontal axis of a rectangular coordinate system, to the left and to the right, respectively. $\sqrt{-1}$ is a vector on the *vertical* axis (upward). That is, the vector $\sqrt{-1}$ is -1 rotated clockwise 90° from the horizontal axis to lie "halfway" between -1 and $+1$. Since -1 is a negation, as is NOT, then associating the square-root (of -1) with the square root (of NOT)—since both do something "half-way"—seems a natural one to make. You can find more on this imagery in my book *An Imaginary Tale: The Story of* $\sqrt{-1}$, Princeton University Press, 2010.

21. G. J. Milburn, "Quantum Optical Fredkin Gate," *Physical Review Letters*, May 1, 1989, pp. 2124–2127. Fredkin gates preserve parity, you'll recall, which means there is no "loss" of photons from input to output. That's good, because to "lose" a photon means its was absorbed, which generates heat, the fatal signature flaw of irreversible computation.

22. C. Munroe et al., "Demonstration of a Fundamental Quantum Logic Gate," *Physical Review Letters*, December 18, 1995, pp. 4714–4718. See also Monroe and David J. Wineland, "Quantum Computing with Ions," *Scientific American*, August 2008, pp. 64–71.

23. For more on time travel paradoxes in general, and bilking paradoxes in particular, see my book *Time Machines*, Springer, 1999, pp. 245–353. The first discussion in the physics literature that I know of, concerning a time travel bilking paradox, was due (perhaps not surprisingly) to Feynman and his Princeton University doctoral dissertation adviser John Wheeler (1911–2008), in a 1949 paper in *Reviews of Modern Physics*. I discuss the details of the Feynman-Wheeler bilking paradox in *Time Machines* on pp. 332–336.

24. David Deutsch, "Quantum Mechanics Near Closed Timelike Lines," *Physical Review D*, November 15, 1991, pp. 3197–3217, which develops a paradox-free way of using time travel for computation. The term "closed timelike lines" (or "curves") is physicist-lingo for *time machine*. Deutsch (born 1953), presently at the University of Oxford, has long been a champion of the "many-worlds" idea (every time a decision is made the entire universe splits to

accommodate every possible outcome), a staple in science fiction long before (some) physicists adopted it; see *Time Machines* (previous note), pp. 214–303. This concept explains, according to Deutsch, the power of a quantum computer. As he writes in his 1997 book *The Fabric of Reality* (p. 217): "To those who still cling to a single-universe world-view, I issue the following challenge: *explain how Shor's algorithm works*. I do not merely mean predict that it will work, which is merely a matter of solving a few uncontroversial equations. I mean provide an explanation. When Shor's algorithm has factorized a number, using 10^{500} or so times the computational resources that can be seen to be present, where was the number factorized? There are only about 10^{80} atoms in the entire visible universe, an utterly minuscule number compared with 10^{500}. So if the visible universe were the extent of physical reality, physical reality would not even remotely contain the resources required to factorize such a large number. Who did factorize it, then? How, and where, was the computation performed?" Many physicists, not surprisingly, find the many-worlds idea of splitting universes outrageous.

25. Dave Bacon, "Quantum Computational Complexity in the Presence of Closed Timelike Curves," *Physical Review* A, September 2004 (available online).

Hint: the matrix for the quantum controlled-NOT (CNOT) gate is

$$\begin{bmatrix} 1 & 0 & 0 & 0 \\ 0 & 1 & 0 & 0 \\ 0 & 0 & 0 & 1 \\ 0 & 0 & 1 & 0 \end{bmatrix}$$

To see this, pre-multiply the input state matrix

$$\begin{bmatrix} 0 & 0 \\ 0 & 1 \\ 1 & 0 \\ 1 & 1 \end{bmatrix}$$

by the quantum CNOT gate matrix and observe the result.

Epilogue

For the Future: The Anti-Amphibological Machine

To end this book on a math-free note, what follows is a personal vision of the sort of logical problem that may soon be one that even a quantum computer would find a struggle to deal with—the decipherment of entangled legalese, the sort of monstrous gobbledegook one finds, for example, in the increasingly convoluted IRS tax code (without TurboTax®, your author would long ago have gone quietly insane trying to file a Federal 1040 that didn't have at least 73 "errors" in it). In the form of a short story, that vision is "The Language Clarifier," which first appeared in the May 1979 issue of *Omni Magazine*. The original title was "The Anti-Amphibological Machine," but *Omni*'s fiction editor thought that a tad too mystifying for readers and "suggested" the change. Now, at last, I can use my original title for the final section of this book.

I think that both Boole and Shannon would have been sympathetic to the story's underpinnings in the law; after all, you'll recall that Boole won his 1858 Keith Prize for a paper applying probability to legal testimony, and Shannon's father was a judge. A "language clarifier" is still an imaginary invention, but I think the world really could use such a gadget. Perhaps a reader will rise to the challenge? Now, two historical notes before the story. First, the hero of my little tale, physics professor Sam Sklansky, got his name in honor of my doctoral dissertation advisor Jack Sklansky, professor emeritus of electrical engineering and computer science at the University of California at Irvine. I hope Jack will forgive me for transplanting him from EE to physics. And second, my sister Kaylyn is married to a trial lawyer (a wonderful guy at weekend barbecues but a shark in the court room), and her husband Kent is definitely not a "fathead."

The Language Clarifier

The idea for the invention came during the divorce. He knew he was going to be screwed, but with the legal mumbo jumbo of the separation agreements, he couldn't figure out *how* he was going to be screwed. Janet's damn fathead lawyer had drawn them up—he'd even given the go-ahead for that, as he hadn't planned to contest her. After all, he *had* been caught in a rather blatant, clear-cut position of adultery. At the time, he had thought the wild-passioned honey-blonde had been worth it, but now he was beginning to have doubts.

He had a doctorate in semantics and was the author of two scholarly tomes on the meaning and structure of words, but Professor Willard Watson still couldn't understand what in hell was going on. Did he or didn't he get to keep the car? How about the house, the savings account, the cat and dog, the antique hutch, the silver, the ski equipment, the home library, the television sets, and all the rest of the earthly possessions collected over twenty-five years of marriage? And what about alimony? Asking Janet's fathead lawyer led merely to the receipt of additional incomprehensible letters, notices, and other horrible documents. Just what the heck *did it mean* to receive a letter saying:

> *Notice is hereby granted to Willard Watson, the first party of aggravation with respect to the aggravated second party, Janet Watson, of an action for divorce, in the County of Orange of the State of California. Actions involved include but may not be fully delimited by their listing here: the exposure of the second party to loathsome disease by the first party due to participation in perverted crimes against the order of nature; public embarrassment of the second party due to the wanton, unrestrained, lascivious behavior of the first party; other acts committed by the first party of various horrid natures, to be specified at a later time, as needed, to describe the untenable position*

> *of the second party with respect to the first party. The second party maintains total freedom in the question of complicity of action, and, except in those cases where litigation proves contrariwise, sues for all common hereditaments, past, present, or future, to revert to the second party, except for the sole ownership of items, things, or other states of being in possession of the first party, prior to the initial date of marriage between the first party and the second party, except for such entities excluded by prior agreements in force at that time, or at other times not mentioned in this action.*

Professor Watson was somewhat perplexed by all this. So he hired his own fathead lawyer.

What Professor Watson ended up with then was *twice* as much paper that he couldn't understand. Willard learned the truth of the old New England saying: "A man between two lawyers is like a fish between two cats." So he fired his fathead lawyer. And he stayed up for three straight nights, mulling over his desperate situation until the idea for the invention came to him. He quickly made an appointment to see his old friend at the college, Professor Sam Sklansky of the Physics Department.

It was a cold, windy, and rainy day in early October as Willard ran from the parking lot to Sklansky's office. His shoes soon filled with water, and he squished his way up the steps into the Physics Building. Even Nature was dumping on him now.

Sklansky's door was open, and he walked in, dripping, sloppy wet, with water slushing out his hat brim onto the floor. "Hi, Sam. Thanks for seeing me so early in the morning." He stood there, looking like a lone, forlorn weed in the middle of a growing pool of water.

Sklansky, a brilliant, very direct sort of fellow, looked quizzically back. "So what's the problem, Willard? And by the way, umbrellas, raincoats, and boots *have* been invented. You some kind of health nut, running around in the rain like nature boy?"

"Look, Sam, I'm desperate, and I've had a lot of things on my mind besides the weather. I need your help, and I need it fast. Janet's going to rake my behind over the coals, but good, if I don't get someone to tell me what the divorce settlement she's serving on me *means*!"

"Willard, you want to see Professor Shyster over in the Law School. I deal in physical facts, mathematical validity, in *cosmic truth*, not the mental hash–mish-mash of lawyers!"

"No, Sam, another fathead lawyer isn't what I need. I need *you*. I want you to tell me if my idea is possible."

So, good friend that he was, Sam listened. At first he laughed hysterically, then he wrote a few Boolean equations, and seeing a little hope, he wrote some more. Then he became quietly excited, and finally, as Willard wrapped up his arguments, Sam became hysterical again, but this time it was with excitement. It *could* be done. The two old friends shook hands and agreed to begin construction that very weekend. Willard would provide the description of the necessary syntactical transformations, along with a complete table look-up dictionary of all the required synonyms, antonyms, and transitive verbs with irregular conjugations. Sam would provide the electronic expertise, produce the wiring schematics, order the parts, and do all the soldering.

It was just two weeks later that they stood in Sam's laboratory, looking at their gleaming creation. A cubical box, precisely 119 centimeters on an edge, it had a smooth, featureless appearance, with the double exception of two horizontal slots. One was marked **INPUT** and the other **OUTPUT**. It was ready for testing.

"Okay, Sam, you designed it, you can have the honor of the first test."

"No, it was your idea, so you go ahead."

"Please, Sam, I insist."

"Well—all right, I *do* just happen to have a test problem ready." So saying, Sam walked over to his desk, rolled a fresh piece of heavy white bond paper into his typewriter, and quickly snapped out in bold pica letters:

> *Liquid precipitation fell from the heights, followed by*
> *the spherical solid version, with the process terminated*
> *by the reverse transport in the gaseous state.*

Sam took the sheet over to the machine, and with an expression that was a mixture of glee and apprehension, held it up to the

INPUT slot. "Ready, Willard?" At the nod of his friend's head, Sam pushed the paper in. After only a few seconds, another piece of paper shot from the **OUTPUT** slot. Both men grabbed it in midair, and together read:

First it rained, then it hailed, and finally the water evaporated.

"Well, I'll be damned!" they exclaimed in unison. The Language Clarifier worked.

"Hey, hey, hey, Sam, it looks good, it looks good!" Willard began to paw through his briefcase, looking for his divorce papers. "*Now* I'll find out just what that scheming wife of mine is up to!"

"Wait, Willard," said Sam, as he placed a restraining hand on his friend's shoulder. "Let's not be hasty. We should really test it some more. Look here, I have a copy of today's campus newspaper carrying an interview with the Undergraduate Dean. Listen to this, will you, the perfect test!" He read aloud:

> *Even in institutions like our college, which may be expected to have rather homogeneous populations, one encounters a tremendous diversity in the family subcultures that students come from, in addition to the idiosyncratic mix of assets and liabilities that characterize them.*

"Wow, Sam—do we dare put *that* into it? It could blow the circuits!"

"Might as well find out if the Language Clarifier *really* works, Willard." Sam soon had the dean's words typed in clear, crisp, sharp letters. He shoved them into the **INPUT** slot, and the machine responded in seconds with:

> *No two students are alike.*

"Son of a gun, Sam, look at that! The translation actually makes sense. Try something else on it."

"Okay, Willard. Take a look at *this*—another quote from the Dean":

> *We thus encounter students whose educational aims are crystal clear, as well as others whose purposes have all the clarity of an amorphous*

> *mist emanating from a thick cloud of existential miasma.*

Quickly they typed this out and inserted it into their machine, and they were soon in possession of the machine's response:

> *Some students know what they want, and others don't.*

"That's enough for me, Sam—it works! Now, where the heck *are* those damn lawyer's papers!"

* * * * * * * *

The rest is history. Willard found out what the divorce was going to cost him. He still got screwed, of course, but with the Language Clarifier deciphering the papers from Janet's fathead lawyer, he knew precisely *how* he was being screwed. Actually, Willard was really unconcerned, as he and Sam expected to make a bundle selling their machine to business, higher education, and government. Their need for extreme clarification was well established. Let Janet have *everything*—secretly, Willard was *happy* to be rid of the damn cat and dog. He would recoup it all, and more, with the royalties from the Clarifier.

Willard let Sam handle the business end of the Language Clarifier, and it was with some greedy anticipation that he dropped in on him after the divorce was settled. Willard was flat broke.

"Okay, Sam, give me the news. How are we doing with the Clarifier?"

Sam opened his desk drawer, pulled out a piece of paper, and handed it across to Willard. It was a cashier's check for fifty thousand dollars. "There you are, Willard, your share of the proceeds from our first three sales. And more to come!"

"Hot damn, Sam. I knew it! Who bought the first three machines—businessmen dealing with government regulatory agencies?"

Sam grinned at Willard. "Professor Shyster, over at the Law School, bought all three."

"Of course," exclaimed Willard, slapping his forehead with a hand. "Lawyers *would* be the prime users of the Clarifier, wouldn't they? Why, with all the ritual chants they produce, they'll be in the market for Clarifiers for the next fifty years. What's old Shyster going to do with them, anyway?"

"Actually, Willard, you've got it backward. Shyster is writing a law book, and he's found that his early drafts weren't really up to par as far as the publisher is concerned. Not scholarly-sounding enough, or something like that. So the Clarifier is just what he needed."

"I don't get it, Sam," said Willard, with a puzzled look on his face. "If Shyster's book isn't impressively complex enough, how's the Clarifier going to help?"

Sam leaned back in his chair with a pleased look on his face. "Willard, my boy, there's an old rule of thumb in physics that says if a process works in one direction, it will almost always be true that it can go the other way, too."

Then Willard understood. "You don't mean, you couldn't *possibly* mean—"

"Yep, that's right. I just moved a couple of wires around, and now old Shyster just stuffs his clearly written book draft into the **OUTPUT** slot, and the most incomprehensible muddle you could possibly imagine emerges from the **INPUT** slot. Should be a legal best-seller. Shyster thinks the Harvard Law dean will soon come calling with the offer of a tenured appointment."

Willard was stunned. The irony of it was mind-boggling. As he stared at Sam, his friend chuckled. "Look at it this way, Willard, how many of the lawyers who'll read it will *really* know, or even give a damn, whether they understand it or not?"

Before Willard could respond, Sam's secretary put her head into the office.

"Excuse me, Professor Sklansky, but this large envelope, from Washington, just came for you registered, special delivery. It looks important, so I thought I should give it to you right away."

"Yes, good, thank you, Susan." As the pretty young lady left, Willard found himself admiring her slender ankles, the motion of her firm thighs under a snug dress, her really spectacular bottom. "Careful, Willard," cautioned Sklansky, the always observant physicist. "As I recall, it was a blonde who did you in last time, and besides, she's the best damned secretary I've ever had. So stay away from her!"

"Ah, I suppose you're right, Sam, but she *is* a nifty-looking gal."

"Hmmph," grunted Sam, who had been reading the just-delivered mail. A slight frown was forming on his mouth. "Listen to *this*, Willard, it's from the Chief Legal Officer of Defense Research and Engineering in the Pentagon. Remember, I wrote to them about the Language Clarifier—pointed out how they could use it to decipher the thousands of proposals they get from industrial contractors every year?"

He read:

> *Replying to your communication of 28 October, we have, after analysis of the broad ramifications of and pertaining to, in all its present and future forms, the Language Clarifier, found it to present a less than superior hold on the financial, economic, reputational, and any other forms of gain, physical or otherwise, of its inventors. In view of the willingness of said inventors to receive and accept a yearly stipend, in perpetuity, or for life, whichever terminates first, of one million dollars, they shall also accept the impact and import of the Military Secrecy Act of 1947, Title 12, Section 19.321 (see attached forms). Return of this document, with said inventors' signatures, will constitute a mutually satisfactory agreement. Otherwise, not, with all applicable consequences to follow (see attached forms).*

Sam put the letter down on his desk and drummed his fingers on the hard wooden surface. "Well, Willard, what do you make of *that*?" He idly flipped through the fifty-three single-spaced onionskin pages of the 1947 Military Secrecy Act. "Frankly, Willard, it sounds to me like the bastards are afraid to have the Clarifier around! You know, if the military boys can use it to blow away the industrial proposal-writer's crap, I suppose industry could use it to dig through all the government's crud, too. Why, *both* sides would have to make sense. Imagine that! And just think of the heart attacks in Congress when all the blowhards on Capitol Hill realized their speeches would be run through the Clarifier!"

"Christ, Sam, how the hell should *I* know? Look, let's run this letter and the Secrecy Act through the Language Clarifier—you still have our prototype unit in your lab, right?"

"Right, Willard. Let's go!"

A few minutes later the **INPUT** slot gobbled up the Pentagon letter. Then the 53 pages of the Military Secrecy Act of 1947 followed. A full forty-three seconds ticked by as the Clarifier mulled over its latest task. Deep in its bowels a few transistors grew hot, an amplifier oscillated violently with feedback but recovered before vaporizing, and a mechanical gear-train drive almost ground off a tooth or two. Then, finally, the Clarifier finished. It ejected its response.

> *Sign the agreement, forget you ever heard of the Language Clarifier, and you get a megabuck a year for life. Don't sign the agreement, and they toss you in the slammer (with one 60-second cold-water, low-pressure shower every 10 days) and throw away the key.*

* * * * * * * *

Sam lives in Hawaii now, retired from teaching, and is writing a book on the physics of hanging ten. Willard quit teaching, too, married Susan, and it would be indelicate to discuss what *they* are doing. Once a year they meet in San Francisco, split the million bucks, have a few drinks at Fisherman's Wharf, and ride the cablecar.

Oh, yes, Sam was right. Old Shyster's book *was* a best-seller, and he's now being hotly pursued by both Harvard *and* Yale Law, thus proving you don't have to be smart to get paid a million bucks a year for forgetting what you know and doing nothing.

Quite often, merely being a fathead lawyer is sufficient.

Appendix

Fundamental Electric Circuit Concepts

0101100101100101100 10

In this brief appendix I'll give you a superquick review of all you need to know—and nothing beyond that!—to understand the electrical circuits in this book.

All of our logic circuits use only resistors. Diodes and relay coils appear, too, but how *they* work is discussed in the text. So, let's start with how a resistor—a component with two terminals — is mathematically defined. The definition will tell us how the current (i) in a resistor (R) is related to the voltage drop (v) across the terminals, with Figure A1 in mind. But of course we are immediately faced now with the questions of what do we mean by *current* and *voltage drop*?

Current is the motion of electric charge, that is, the motion of electrons, subatomic particles each of which possess the negative electric charge of $1.6 \cdot 10^{-19}$ coulombs (named after the French physicist Charles-Augustin de Coulomb (1736–1806)). The current i at any point in a circuit is defined to be the *rate* at which *positive* charge moves past that point; 1 ampere of current is equal to 1 coulomb per second. Since electrons carry negative charge, their motion is equivalent to positive charge moving in the opposite direction; that is, the actual physical motion of the electrons is *opposite* to the direction of i. The *ampere* is named after the French physicist André Marie Ampère (1775–1836), who showed that an electric current generates magnetic effects.

Voltage is defined as the energy per unit charge; a common source of voltage is the ordinary 1.5 volt battery. The **voltage drop** (from plus to minus) across a resistor is the energy *expended* (appearing as heat) in transporting a unit charge through the resistor. If a battery is connected to a network of resistors, the electrons at the negative terminal of the battery move through that network and return to the

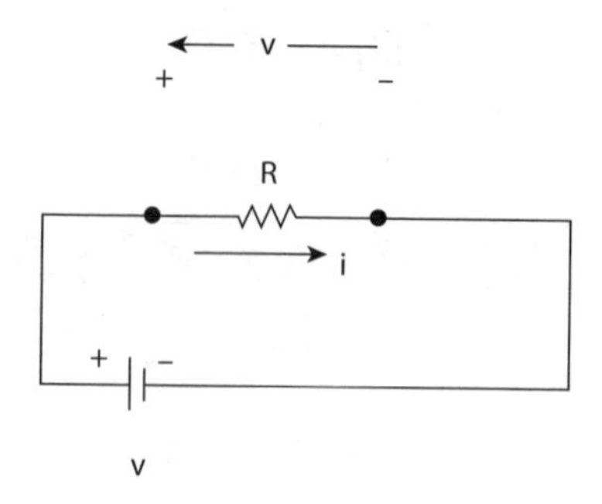

Figure A1. The resistor.

positive terminal of the battery, and there is a 1.5 volt drop across the network. The *volt* is named after the Italian physicist Alessandro Volta (1745–1827), who constructed the first battery in 1800.

Now we can define how resistors work. They obey *Ohm's law*, namely,

$$v = iR,$$

where R is measured in ohms (named after the German physicist Georg Ohm [1780–1854], and, again, take a look at Figure A1, where the symbol for our voltage source (a battery) is the standard one of two parallel lines (the long line is the positive terminal and the short line is the negative terminal).

In the analysis of resistor circuits, two incredibly useful laws are used, called *Kirchhoff's laws*, after the German physicist Gustav Robert Kirchhoff (1824–1887). They are actually the fundamental conservation laws of energy and electric charge. In words:

> **Kirchhoff's voltage law**: the sum of the voltage drops around a closed path in any circuit is always zero (this is true in *any* circuit, not just resistor circuits). This physically says that the net energy *change* for a unit charge that travels completely around a closed path is zero. If it were not zero, then we could repeatedly transport the charge around the closed path in the direction in which the net energy change is positive (there are of course *two* ways to go around a closed path; and if the change isn't zero, then one way it will be positive and the other way it will be negative) and so we could become rich selling the excess energy to the

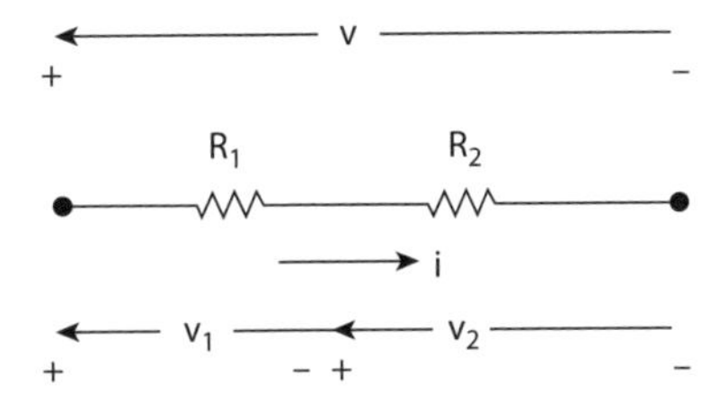

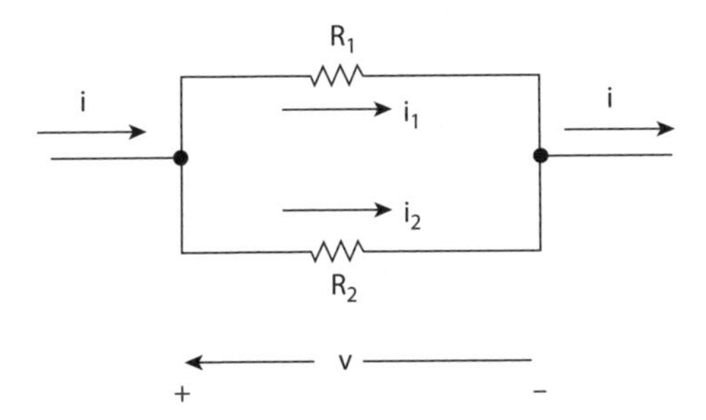

Figure A2. Resistors in series (top) and in parallel (bottom).

local power company! Conservation of energy says we can't do this (recall perpetual motion machines).

Kirchhoff's current law: the sum of all the currents into any point in any circuit is always zero (this is true in *any* circuit, not just resistor circuits). If this weren't so, then at the point there must at each instant be either charge being created or being destroyed, which the conservation of electric charge denies. Whatever charge is transported "into" a point by some currents must be transported "out" by other currents.

To see how useful these two laws can be, let's derive the rules for how two resistors (R_1 and R_2) combine when connected in *series* and in *parallel*, as illustrated in Figure A2. When in series, the two resistors carry the same current i, but in general have different voltage drops (v_1 and v_2). When in parallel, the two resistors have the same voltage drop v, but in general have different currents (i_1 and i_2). In each case, let's write R_e for the single equivalent resistor that replaces the two original resistors.

In the series case, we have

$$v = iR_e,$$

$$v_1 = iR_1,$$

$$v_2 = iR_2,$$

and

$$v = v_1 + v_2,$$

where the last equation follows from Kirchhoff's voltage law. So,

$$v = iR_e = iR_1 + iR_2,$$

and our result is that

$$R_e = R_1 + R_2.$$

In the parallel case, we have

$$v = iR_e,$$

$$v = i_1 R_1,$$

$$v = i_2 R_2,$$

and

$$i = i_1 + i_2,$$

where the last equation follows from Kirchhoff's current law. So,

$$i_1 + i_2 = \frac{v}{R_1} + \frac{v}{R_2} = i = \frac{v}{R_e}$$

or,

$$\frac{1}{R_e} = \frac{1}{R_1} + \frac{1}{R_2}$$

and our result is that

$$R_e = \frac{R_1 R_2}{R_1 + R_2}.$$

Acknowledgments

This is the one page, when creating a book, that I always look forward to writing. Besides meaning I am at last done, this is where I get to thank all the many people who have shared with me the substantial effort required to write a technical book. The idea for this book came during a dinner conversation a few years ago with John Pokoski, a fellow emeritus professor of electrical engineering at the University of New Hampshire. John, whose early career at IBM paralled mine as a digital circuit logic designer in the 1960s, told me he had often flipped through many of my books at libraries and bookstores, but hadn't read any of them. If, however, I should ever write a book on Boole, *that* one he promised he *would* read. The challenge was irresistible, and that's how this book came to be.

The talented people at Princeton University Press, nearly all of whom I have had the pleasure of working with on past books, were central to the production of this book. Specifically, my terrific editor Vickie Kearn and her equally professional colleagues Stefani Wexler, Dimitri Karetnikov, Alison Anuzis, Quinn Fustin, Erin Suydam, Carmina Alverez-Gaffin, and Debbie Tegarden. I received very helpful feedback on the book from two academic physicists who reviewed the original typescript for Princeton: Lawrence Weinstein at Old Dominion University in Virginia, and Charles Adler at St. Mary's College of Maryland.

The book's copyeditor, Alice Calaprice (who is a well-known author in her own right, and a former senior editor at Princeton) was a pleasure with whom to work. Alice saved me from more than a few missteps.

At the MIT Museum (Cambridge, MA) curatorial assistant Ariel Weinberg was of great help in obtaining the photo of Claude Shannon, while at University College (Cork, Ireland) archivist Carol Quinn provided gracious support in my quest for a photo of George Boole. Artist Randy Glasbergen allowed me to use his very funny penguin cartoon in a fashion he almost certainly didn't have in mind when he drew it.

MATLAB® is a registered trademark of The Math Works Inc. and is used with permission. The MathWorks does not warrant the accuracy of the text in this book. This book's use of a MATLAB® related products does not constitute an endorsement or sponsorship by The MathWorks of a particular pedagogical approach or particular use of MATLAB® software.

And finally, I thank my wife Patricia for providing me with the fifty years of emotional support (combined with lots of common sense!) that have made my writing-life possible. She has been a behind-the-scenes co-author on all of my books in a most important way. Our two tiger-tabby cats (Vixen and Tigger) also deserve some mention as well, as they greatly reduced the stress of writing when they *finally* learned that they absolutely must *not* attempt to eat the book's electronic files stored on my computer's flash-drives with their seductively pretty (to a cat), flashing lights.

Index

Also by Paul J. Nahin

Oliver Heaviside (1988, 2002)

Time Machines (1993, 1999)

The Science of Radio (1996, 2001)

An Imaginary Tale (1998, 2007)

Duelling Idiots (2000, 2002)

When Least Is Best (2004, 2007)

Dr. Euler's Fabulous Formula (2006, 2011)

Chases and Escapes (2007)

Digital Dice (2008)

Mrs. Perkins's Electric Quilt (2009)

Time Travel (2011)

Number-Crunching (2011)